REVISION QUESTIONS ON TABLE AND BAR

Jeffrey T. Clarke

Senior lecturer, Worcester Technical College

To accompany TABLE AND BAR: A guide to alcoholic beverages, sales and service

Edward Arnold

A division of Hodder & Stoughton

LONDON   NEW YORK   MELBOURNE   AUCKLAND

© 1989 Jeffrey T. Clarke

First published in Great Britain 1989

British Cataloguing in Publication Data

Clarke, Jeffrey T.
  Revision questions on table and bar
  1. Alcoholic drinks —— For bartenders
  I. Title
  641.8'74

  ISBN 0–7131–7816–7

Typeset in Singapore by Colset Private Limited.
Printed and bound in Great Britain for Edward Arnold, the
educational, academic and medical publishing division of Hodder and
Stoughton Limited, 41 Bedford Square, London WC1B 3DQ by
Richard Clay plc, Bungay, Suffolk

Contents

Preface

This book, the companion volume to *Table and Bar: a guide to alcoholic beverages, sales and service* will enable enthusiastic students of alcoholic beverages to work at their own pace, beginning with the simplest sections and, as the course progresses and knowledge increases, moving on to the more difficult parts.

Attempt the work honestly, marking conscientiously, and concentrating your studies on those areas where you find yourself to be weak.

About 95 per cent of the answers can be found in *Table and Bar*, so by all means use it to look up the answers if you wish, as this is in itself an excellent way of learning. If you are more confident, you may find satisfaction in attempting the whole series of questions without using the textbook. Answers to all the questions are given at the end of the book.

I hope that you will find *Revision Questions on Table and Bar* to be a fun way of learning about a subject which has given me, and I am sure will give you, lifelong pleasure.

Jeffrey T. Clarke

Section one
Wine from many countries

1 French wine and spirit regions

The sixteen French wine and spirit producing regions are, in alphabetical order: Alsace, Armagnac, Bordeaux, Calvados, Chablis, Champagne, Cognac, Jura, Loire, Midi, North Burgundy, North Rhône, Provence, Savoie, South Burgundy and South Rhône.

Match these to the numbered areas of the map, writing the answers in the spaces provided. Score one point for each correct answer.

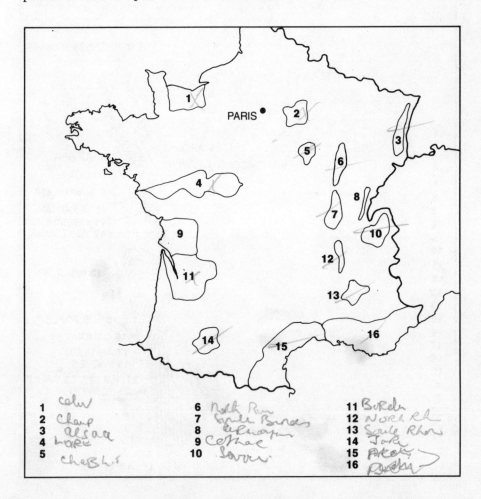

PARIS

1
2
3
4
5
6
7
8
9
10
11
12
13
14
15
16

1 Calv
2 Champ
3 Alsace
4 Loire
5 Chablis
6 North Rhn
7 South Burdes
8 Champagne
9 Cognac
10 Savoie
11 Bordeaux
12 North Rh
13 South Rhon
14 Jura
15 Prconce
16 Provence

2 *Wine regions of Burgundy*

On this map of Burgundy 25 place names have been jumbled.

Unscramble them, writing the correct name in the numbered space provided. Score one point for each correct answer.

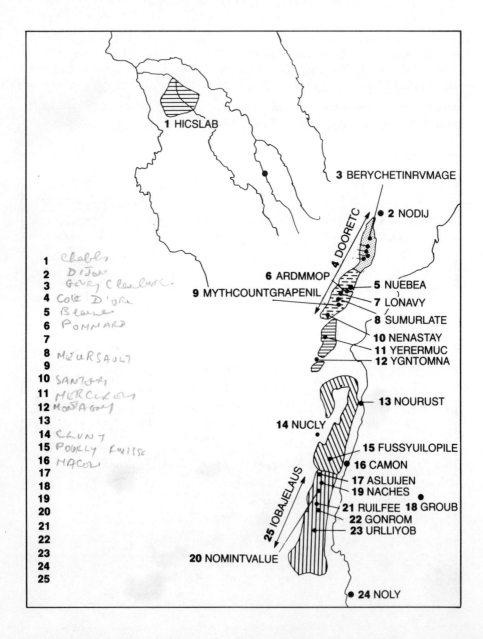

1 Chablis
2 Dijon
3 Gevry Chamber...
4 Côte D'Or...
5 Blanc
6 POMMARD
7
8 MEURSAULT
9
10 SANTAY
11 MERCULEY
12 MONTAGNY
13
14 CLUNY
15 POUILLY FUISSE
16 MACON
17
18
19
20
21
22
23
24
25

1 HICSLAB

3 BERYCHETINRVMAGE

2 NODIJ

4 DOORETC

6 ARDMMOP

9 MYTHCOUNTGRAPENIL

5 NUEBEA

7 LONAVY

8 SUMURLATE

10 NENASTAY

11 YERERMUC

12 YGNTOMNA

13 NOURUST

14 NUCLY

15 FUSSYUILOPILE

16 CAMON

17 ASLUIJEN

19 NACHES

25 IOBAJELAUS

21 RUILFEE 18 GROUB

22 GONROM

23 URLLIYOB

20 NOMINTVALUE

24 NOLY

Points possible 25   Points obtained _____

3 *French wine*

1 Red wine from Bordeaux is known by the name _____

2 *Botrytis cinerea* is a mouldy condition of grapes which produces the fine sweet white wines of Bordeaux in the district of _____

3 The land between the rivers Dordogne and Garonne which produces dry white wines is _____

4 The region of Bordeaux belonged to England because of the marriage of Henry (Plantagenet) to _____

5 The area west of the river Gironde which produces some of the world's finest red wines is the _____

6 Name one of the *premiers grands crus classés* châteaux of the 1855 classification of the Médoc _____

7 Burgundy vineyards tend to be small due to a historical event between 1789 and 1791, the _____ (two words)

8 In 1443 Nicolas Rolin founded a hospital for charity called the _____. Funds are raised each year when an _____ is held on the third Sunday in _____ (month) (three points)

9 *Vin de l'année* is produced each year in _____ (area). It is then rushed to England in a race to be first to supply the wine to customers

10 Passe-tout-grains is a wine blend of one-third _____ and two-thirds _____ (two points)

11 It comes from the region of _____

12 Chablis wine would usually be described simply as _____ (colour) and _____ (taste) (two points)

13 The most famous dry rosé wine from the Rhône valley is _____

14 Two other plants are required by tradition to be grown amongst the vines of Châteauneuf-du-Pape. They are _____ and _____ (two points)

15 Côte Rôtie in the northern Rhône valley produces fine _____ (colour) wines

16 Alsace vineyards are protected by the _____ mountains

17 The area of Alsace derived its name from the river _____

18 Alsace wines made from a blend of grapes which are all of 'noble' origin are labelled _____

19 Fine sparkling white wine is made at _____ in the Touraine region of the Loire

20 Muscadet can be described as a _____ white wine

21 Bandol, Cassis, Palette and Bellet are all AOC wines from the region of _____ in south-east France

22 Blanquette de Limoux is a _____ white wine

23 Louis Pasteur lived and worked at Arbois in the region of _____

24 Chinon is a light red wine from the region of _____

25 By law, all Alsace AOC wine can only be _____ in Alsace

26 A chapel was built on a hill by a returning crusader on what are now the Rhône vineyards of _____

27 Beaujolais wines are light red in colour and are made from the juice of the _____ vine

28 Two grape varieties which are grown in North Burgundy as well as Champagne are the _____ and the _____ (two points)

29 Wines which include the _____ grape are expected to keep and mature for many years (eg many clarets)

30 The _____ is a great wind which blows down the Rhône valley and causes problems for the *vignerons*

31 The *appellation contrôlée* laws of France were based on those which Baron Le Roy de Boiseaumarie made for the vineyards at _____

32 The Côte d'Or is the name given to the two French wine districts of _____ and _____ (two points)

33 The English forces lost Bordeaux in 1453 at the battle of _____

34 Popular rosé wines are made in the Loire valley in the district of _____

35 Vin fou is a sparkling wine produced in the region of _____

36 Vin gris is the name given to a wine which is _____ in colour

Points possible 43   Points obtained ____

4 *German wine regions*

The 11 German wine producing regions are, in alphabetical order: Ahr, Baden, Franconia, Hessische Bergstrasse, Mittel-Rhein, Mosel-Saar-Ruwer, Nahe, Rheingau, Rheinhessen, Rheinpfalz and Württemburg.

Match these to the numbered areas of the map, writing the answers in the spaces provided. Score one point for each correct answer.

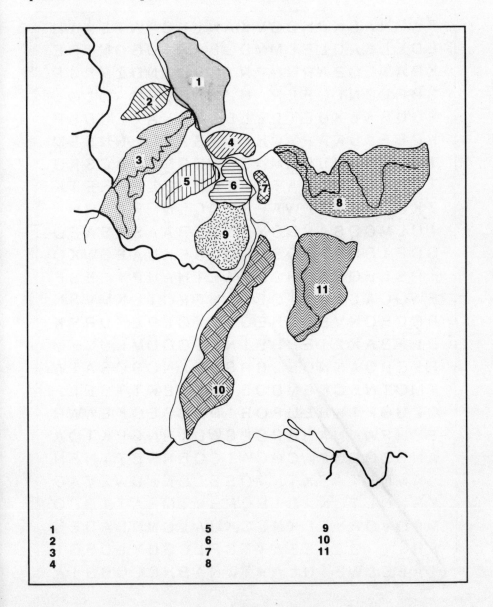

1	5	9
2	6	10
3	7	11
4	8	

Points possible 11   Points obtained ____

5 *German wines and spirits*

There are 25 German words, all to do with wines or spirits, hidden among the letters below, either vertically, horizontally or diagonally (from top to bottom or from bottom to top).

Find and circle them. Score one point for each correct answer.

```
F A G N A L K P N B O V K A N L B G P W T E A I H
E D E L F A U L E L M M D G B A R I U G G N I R E
K P A I E O B A R N G P N I G O D M N O Z M K J F
S B R E B N L Q F L R I H Z J F B U E G A Q L O L
P G D B W L X U E T L L L D G H K B A I E C D S P
R C F F P D K A F S C H L O S S X Y U E N N S E M
I O R R W B U L E M M W E W W O R M S P A C S K G
T O P A L A T I N A T E H T B R Q E L F P E E T K
Z K S U K B R T L W A P K M Q C L N M R L U G H R
I U L M C D S A V D S Y F T P S R A A S X S W F O
G C F I S R R T H Q S N R O U U P K U A E G W K O
I P S L E Q N S S R C M K A E T H A U P T L E S E
B W G C A L T W L T O E N A N F E O T L K M V S N
R C C H O N V E Y A R E G E Y X G L R L P U R V K
D L F B A K X I E H R E I N V U G O O M E U U F M
P F G R O A N N C E L B R S U H L N C S C S A T W
T M O T N B C R E M D O E J W L I B K T T S E L F
Q E U G P I K B L U F Q R I N E C S E O K E W W R
E S H P W N N A H E P O P S W D I R N F P K T O A
A N B Z G E O P W C G C A T C Q R N S B T T N F N
L A M N K T A S X T N M O S E L G N W U W X V A C
K R U C L T Q N A B I R Q R B V T M R A T T O P O
M P O V O A Y E P K N L C K U A L C N N B A D E N
F R L I J O Z R E S W T T B F L C G U Y L D F Q I
T A F E L W E I N L A K T Z N A B H D C Q B S I A
```

6 *Italian wine regions*

On this map of Italy, 26 wine and place names have been jumbled.

Unscramble them, writing the correct name in the numbered space provided. Score one point for each correct answer.

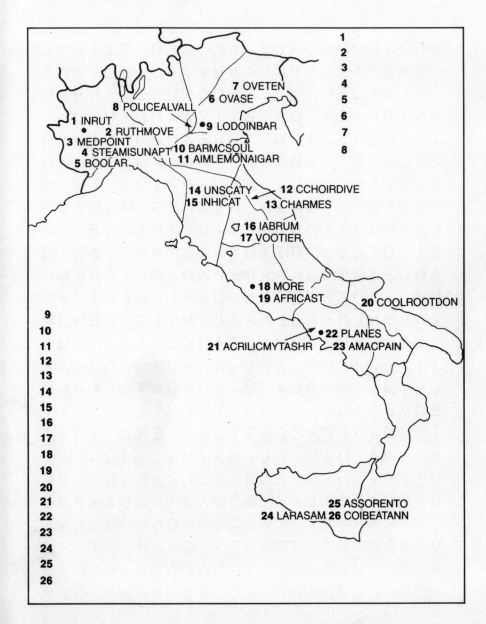

On the map:

- **7** OVETEN
- **6** OVASE
- **8** POLICEALVALL
- **9** LODOINBAR
- **1** INRUT
- **2** RUTHMOVE
- **3** MEDPOINT
- **10** BARMCSOUL
- **4** STEAMISUNAPT
- **11** AIMLEMONAIGAR
- **5** BOOLAR
- **14** UNSCATY
- **12** CCHOIRDIVE
- **15** INHICAT
- **13** CHARMES
- **16** IABRUM
- **17** VOOTIER
- **18** MORE
- **19** AFRICAST
- **20** COOLROOTDON
- **22** PLANES
- **21** ACRILICMYTASHR
- **23** AMACPAIN
- **25** ASSORENTO
- **24** LARASAM **26** COIBEATANN

Numbered list:

1
2
3
4
5
6
7
8
9
10
11
12
13
14
15
16
17
18
19
20
21
22
23
24
25
26

Points possible 26   Points obtained ____

7 *Italian wine*

There are 25 Italian words, to do with wine, hidden among the letters below either vertically, horizontally or diagonally (from top to bottom or from bottom to top).

Find and circle them. Score one point for each correct answer.

```
P C T T C M U M D A K L B K A G S F R I Z Z A N T E
O B K B B A Y U E Y N E A S H Z J B G I W O L P W T
I M L A K C E G G O V E R N O B P U R D L A N R J N
W N T R U A E I L D B M B I I V W A N O F O Y C L A
I B G D W S S K K G R L A B R E A U R M R H A B V R
L O C O R O T O N D O B R K I W N A K D O V G U J O
X T M L Y V E S T B J A E T N E B B I O L O I H A S
S G A I B I S T C W A R S B V Z N Z B Y W D M E S S
L T T N S D T I D H O M C V A U A T H Z A N E T T O
B I U O T X E J J D I X O N L L G T B P K V W N I O
A G D B X Z S O B K A A N T P W O S O A V E R N S C
M N L F P O T F I O D T N G O D F L P H E R T J P W
A S B M N L D B E O L N S T L F H I I O T D O M U M
B L A C R Y M A C H R I S T I P C Q A A Y I N B M I
I I F U R J F R P Z N R V L C X A U Y S D C U C A E
L N Y N L O M B R G A T M V E U F S R F C C K R N C
E O M U C I G E L K C U D U L E R M F L R H U W T Y
T C M N J J F R A S C A T I L A E I E N D I I D E A
A T A W A L U A F Y R L L H A T M M L I S O C L W I
D I C F V T H N O P E O A R E Z L B K Q B R C D B E
B A E M A R S A L A K O W S O A B X R O B A R F N D
T Y J R W N V A Z V C W Q C S E M D Q U P M W B N A
U F B G P I B T E C P N U T K I C N I B S R I L T S
P J P V Y V A E E X I R J W N W C W E S L C N M S C
F O G C L E R S A O C I A O C C X O L A N P O O P P
```

8 *Champagne and sparkling wine*

1 A double bottle of champagne is called a _____

2 The main reason for the superior quality of champagne is probably the _____ soil

3 The three classic grape varieties used in the making of champagne are:
(a) _____ (b) _____ (c) _____
(three points)

4 The monk who is generally thought to have invented champagne was _____

5 More than 75 per cent of the grapes used to make champagne are _____ (colour)

6 The abbey of _____ is where champagne was first made to sparkle

7 To make pink champagne, red wine is added from the village of _____

8 Bubbles were often stirred out of champagne using a _____

9 The two river valleys where champagne is made are the _____ and the _____ (two points)

10 Champagne from all white grapes is known as _____

11 How many *grande marque* champagne houses are there? _____

12 The press for champagne grapes holds _____ kilos of grapes approximately

13 The first pressing of champagne grapes gives ten casks of _____ (2000 litres in all)

14 The fourth pressing gives one cask of _____

15 The blending together of the wines of the different villages in Champagne is known as _____

16 _____ is the daily methodical shaking of the champagne bottles to remove the sediment

17 The final removal of the sediment from the champagne bottle is called _____

18 Topping up with cane sugar in solution in similar wine is known by the term _____

19 Very dry champagne would probably be labelled _____ or _____ (two points)

20 The word on the label of the champagne bottle indicating that the wine is very sweet is _____ or _____ (two points)

21 The best shape of glass for the service of champagne is a _____

22 A jeroboam is equal to _____ bottles of champagne

23 *Cuve close* is a method of secondary fermentation in a closed

_____

24 When removing the cork from a bottle of sparkling wine, hold the bottle at an angle of about _____° (see Part 3 of 'Table and Bar')

25 Sparkling wines should be served at a temperature of _____°C

26 The cheapest way of making sparkling wine is by the _____ method

27 Two sparkling wines from the Loire valley are _____ and _____ (two points)

28 _____ is a sparkling wine made in 34 communes in the area around Limoux in the south of France

29 Sekt is a sparkling wine made in _____ (country)

30 Name a sparkling wine from northern Italy _____

Points possible 36   Points obtained _____

9 *Spanish wine*

There are 26 words (to do with Spanish wines, regions or terms) hidden among the letters below, either vertically, horizontally or diagonally (from top to bottom or from bottom to top).

Find and circle them. Score one point for each correct answer.

```
C L A J G B I F O K E L D P C P A M C J O I K A H
M G P O I K L G B P O W S U O L P G D P A H J D O
F O S P B C A N E R I E D K L O E N L O N P E O B
L E L A M A N C H A F L H I W C M I K S L E R K L
K P M L N J G S K O F R T B N S E L G B R P E M N
B E N O S S C E D I L N O A F L E T O H I J Z G C
O G I M F C A T A L O N I A Z B Y C E W A D D D H
E R E I J B O D R M R N C P H O I P R P C W E S B
S F P N G C F B U T T D B L E W B N Y G S J L P L
L A S O C K B C D R V L J V M A J B O D E G A O G
M H U R I F M H D I N C U Y D W Z C E P A R F I A
F J B O S T I N T O F I Y B H W T I N A J A R S H
P U N G A F Y K S J I P D W N Y M W R F W P O K J
L N C Y E O C D E A K Q P E N E D E S X D W N G S
G I L R T L O L I C O C W D N P L B B O I E T H O
O D C O F R I R W I T Q N H R O M R R A M G E T M
L H O P S A G D K B O F O M S E Y O J P N B R I B
M A N Z A N I L L A T N M A L A G A F W E A A K A
N D S E A K L R B M I L O L M L S C R D P Q C G L
I P E S P U M O S O S I S E P B H E H S K C H S O
A L C O F N B R C A P D C L H J A W E L S O L A M
M H H P C M S N O G C E A L E O G F K E M D G J Q
E M A I G L A W F N B S T A R R A G O N A G I O R
K G M B F L K C E O J A E O G C O K W I T A D M J
C L E O B H E I A I F H L T D M F A G D L G S C T
```

10 *Wine, other than French*

1 The real meaning of the word *vintage* is _____

2 The wine-making belts around the earth lie between the latitudes _____ ° and _____ ° north and south of the equator (two points)

3 The bitter substance found in stalks of grapes which gives wine its keeping quality is _____

4 _____ sprays are used to prevent black-rot and downy mildew

5 The first modern commercial vineyard in the UK was planted by Guy Salisbury-Jones at _____ in 1951

6 'Everyone bows to tokay' is a saying which started because the wine was hidden in the _____ in Hungary

7 Zinfandel is a red varietal wine from _____

8 The disease which killed virtually all the vineyards of Europe and other parts of the world at the end of the last century is called _____

9 This disease is kept under control by _____

10 James Busby was instrumental in developing the vine-growing techniques in _____ (country)

11 The organisation which controls the wine industry of South Africa is the _____ (initials)

12 The religious group who arrived in South Africa from France and improved wine production were the _____

13 The formation of crystals of tartar in wine after bottling can be prevented by _____ in large tanks

14 The slightly sparkling wine of the Minho district of north Portugal is called _____

15 Because of the large number of table grapes produced, it is true to say that _____ produces more grapes than any other European country

16 *Tinto* is the Spanish word for _____

17 Almost half of Spain's table wine is produced in the region of _____

18 Spanish wine with citrus juice is called _____

19 The Italian wine made from grapes grown on the slopes of Mount Vesuvius is _____

20 Name a dry white wine from eastern Italy _____

21 Some chiantis are improved by the _____ system which involves adding juice from semi-dried grapes about three months after the main harvest

22 The popular emblem for chianti classico is a _____ (two words, two points)

23 Most Romans who wish to drink a dry white wine would probably choose a local
_____

24 German wine made from individually picked late-harvested grapes is called
_____

25 The wine of Franconia in its distinctive bottles is known by the name
_____

26 Liebfraumilch originated from the cathedral city of _____

27 *Spritzig* is the German word for _____

28 Translate 'red wine' into German _____

29 The most widely grown grape variety in Germany is the _____

30 Which German river is famous for its red wine production?
_____

31 The German region of Rheinpfalz is also known as _____

11 *Wine (from many countries)*

Consider the statements given below and indicate by circling which are true and which are false.

1 White wine can be made from black-skinned grapes   TRUE/FALSE
2 Italy and France together produce three-quarters of the
   world's wine   TRUE/FALSE
3 *Phylloxera* can be prevented by spraying   TRUE/FALSE
4 White champagne grapes grow mainly in the south of the
   region   TRUE/FALSE
5 A magnum is equal in quantity to four bottles   TRUE/FALSE
6 The wines of Médoc are made from 'noble rot' grapes   TRUE/FALSE
7 Château Lafite-Rothschild was promoted to 'first growth' in
   1973   TRUE/FALSE
8 Burgundy vineyards are small as a result of peasant unrest   TRUE/FALSE
9 The Hospices de Beaune auction is held in January each year   TRUE/FALSE
10 Beaujolais wines will normally keep longer than Médoc wines   TRUE/FALSE
11 Pouilly-Fuissé is further north than Gevrey-Chambertin   TRUE/FALSE
12 Lavender and thyme grow amongst the vines at Château
   Latour   TRUE/FALSE
13 The vineyards of Hermitage are steeply terraced   TRUE/FALSE
14 Alsace is a German wine–producing region   TRUE/FALSE
15 Muscadet would be a suitable wine with shellfish   TRUE/FALSE
16 German sparkling wine is called sekt   TRUE/FALSE
17 A bottle of steinwein is tall and slender   TRUE/FALSE
18 Chianti wines may be red or white   TRUE/FALSE
19 Rioja produces annually half of Spain's wine   TRUE/FALSE
20 Grao vasco is the best–known wine from Dão   TRUE/FALSE
21 Tartrate crystals in wine are a hazard to health   TRUE/FALSE
22 The South African Steen vine variety is Europe's Chenin Blanc   TRUE/FALSE
23 The Murray valley is an Australian wine-producing region   TRUE/FALSE
24 Sultana is a Californian vine variety   TRUE/FALSE
25 Tiger milk is a Hungarian wine   TRUE/FALSE

Points possible 25   Points obtained ____

12 *Colour of wine*

Place the wines listed below in one of the following categories according to the colour of wine you would associate with the name: (a) red (b) white (c) rosé (d) both.

1 Soave	____	21 Muscadet	____
2 Médoc	____	22 Beaujolais	____
3 Chianti	____	23 Bourgueil	____
4 Tavel	____	24 Fitou	____
5 Rioja	____	25 Piesporter Michelsberg	____
6 Vin gris	____	26 Valpolicella	____
7 Nuits St Georges	____	27 Côte Rôtie	____
8 Retsina	____	28 Frascati	____
9 Tokay	____	29 Barbaresco	____
10 St Emilion	____	30 Dão	____
11 Bardolino	____	31 Anjou	____
12 Entre-deux-Mers	____	32 Châteauneuf-du-Pape	____
13 Montana Cabernet Sauvignon	____	33 Château Musar	____
14 Bull's blood	____	34 Chablis	____
15 Pouilly-Fuissé	____	35 Est, Est, Est	____
16 Setúbal	____	36 Pecs	____
17 Château Simone	____	37 Gevrey-Chambertin	____
18 Sauternes	____	38 Tiger milk	____
19 Orvieto	____	39 Sancerre	____
20 Zinfandel	____	40 Barolo	____

Points possible 40   Points obtained ____

13 *Annual production*

The approximate annual production in millions of gallons is shown on the diagram below for the following countries and areas: California, France, Germany, Italy, Portugal, Spain and USSR.

Write the names of the countries over the appropriate bars, given that California produces 550 million gallons. Score one point for each correct answer.

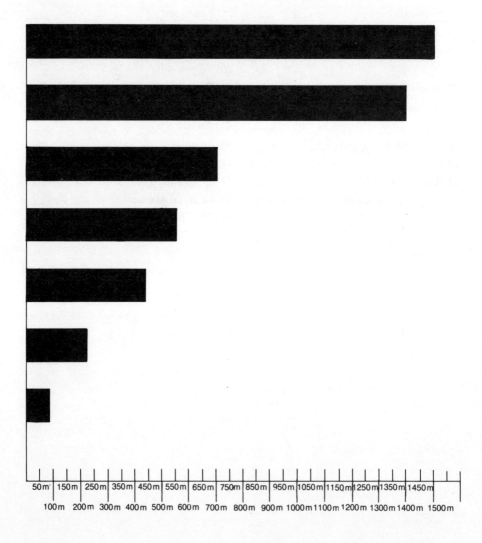

Note As annual wine production varies according to seasonal climatic conditions, figures are approximate mean averages. Statistics for the USSR are estimated.

Points possible 6   Points obtained ____

Section two
The service of wine

1 Case studies on table service

Samantha and Joanne are waitresses at the Far Forest Restaurant, and between them they have increased the trade due to their efficient, speedy and cheerful service. Stephen, the sommelier, is the third of the trio on whom Mr Steele, the manager, largely relies. Most of the other service staff are part-time or family.

Whenever the pressure is on at work the three of them are happiest. They help the slower members of staff and co-operate fully when asked to put in extra time. They have untold patience with difficult customers and have the important knack of making every diner feel important.

The incidents set out below all occurred within the last year and you should carefully consider what the staff's (and your own) reaction would be in these circumstances. Write your thoughts in the space provided, scoring four points for a correct answer. *Parts 3, 5, 10 of 'Table and Bar' should help you here.*

1 A party of six guests had arrived and were choosing food from the *à la carte* menu when Joanne, who was taking the order, noticed that the host had produced two bottles of Greek wine from a small hold-all. What do you think happened during the next few minutes?

2 Stephen presented the wine-list to the Australian couple at the table in the bay-window. When he returned to take the order, Stephen was asked for a red Côtes du Rhône wine to accompany the fish course. What did he do?

3 One Saturday night Mr Steele was helping out as he often did when the restaurant was busy. He had just served a bottle of burgundy red wine to a party of four guests. The host complained that the wine was too cold, so Mr Steele told Stephen to put the wine in the microwave for three minutes. What was Stephen's reaction?

4 On St Valentine's Day a party of 12 people came into the Far Forest Restaurant for dinner. They asked for a bottle of hock to be served with the main course. What did Stephen do?

5 Last Thursday a couple ordered spaghetti bolognese and a flask of chianti. When Stephen served the wine the man sampled it and proclaimed that it was out of condition. What action do you think the sommelier took?

6 The newly married couple on Samantha's station ordered a bottle of Moët & Chandon champagne. When it was being opened they asked especially for it to 'pop'. Did Stephen comply?

7 One of the most upsetting occurrences for Joanne was when a man ordered a bottle of expensive claret with his meal, and when he came to pay he appeared agitated and said he had forgotten his wallet. What action was taken? (*Part 10 of 'Table and Bar' will help here*)

8 A local well-known personality was leaving the restaurant and claimed that his expensive overcoat had been stolen from the coat-hooks in the hall. He said that an old one had been left in its place. He had been served by Samantha.

9 The wine chosen by Commodore Frost was a very old, very expensive vintage red wine which had been in the cellar for many years. It was covered with dust and cobwebs. Stephen sent a *commis* down to bring it up from the cellar. How would it be served?

10 Professor Young, a resident, was dining alone. He looked unwell but nevertheless ordered his customary bottle of white burgundy. After the starter course and having consumed only a half glass of wine, he left the table without a word and went to his bedroom. What did the restaurant staff do?

Points possible 40   Points obtained _____

2 *Glassware recognition*

1 *Beer glasses*   Each of the following glasses is illustrated below. Place the correct letter near the appropriate glass: (a) Pilsner (b) Paris (c) Sleeve (d) Continental (e) Tulip (f) Dimple (g) Wellington (h) Nonik (i) Worthington

2 *Spirit glasses*   As above, place the letter near the appropriate glass: (a) Slim-jim (b) Flanged (c) Brandy balloon (d) Georgian (e) Napoleon (f) Jean (g) Old-fashioned (h) Paris

3 *Sherry and port glasses*   Place the letter near the appropriate glass: (a) Schooner (b) Viking (c) Paris (d) Copita (e) Elgin (f) Club

4 *Table wine glasses*   Place the letter near the appropriate glass: (a) Hock (b) Flute (c) Paris goblet (d) Moselle or Alsace (e) Club goblet

5 *Liqueur glasses*   Place the letter near the appropriate glass: (a) Jean (b) Thistle (c) Elgin

6 *Decanters*   Place the letter near the appropriate decanter: (a) Spirit (b) Ship's (c) Wine

Points possible 34   Points obtained ____

Section three
Fortified wine

1 Fortified wine from many countries A

1 Wines are fortified when they have _____ added
2 The port wine which stays longest in cask is _____
3 Sweet red fortified wine from Catalonia in north-east Spain is called _____
4 Marsala was first produced by two English brothers by the name of _____ in the island of _____ (two points)
5 The driest aperitif madeira is _____
6 The blending system for sherry is known as the _____ system
7 Two types of port which need decanting are _____ and _____ (two points)
8 A cask of sherry containing 108 gallons is called a _____
9 The vineyards where port wine is made in northern Portugal are called _____
10 The main sherry grape variety is the _____
11 Pineau de Charentes is a fortified wine from the _____ region of France
12 Californian angelica is made from the black _____ grape
13 French vermouth is traditionally _____ and _____ (two points)
14 The first modern commercial vermouth was made in Turin in 1786 by _____
15 Madeira wines are heated by a system known as the _____ system
16 A pipe of port contains _____ gallons
17 Commandaria St John is a lightly fortified brown dessert wine from _____ (country)
18 Vermouth takes its name from the German translation of the shrub _____
19 Rainwater is the name of a fortified wine from _____ (country)
20 The yeast crust which forms on the surface of some casks of fermenting sherry is called _____
21 Port wine is taken down the river _____ by rail or by road-tanker in the month of _____ to the town of _____ (three points)

22 The fertile soil of Madeira is _____ and its origins date back to a severe _____ which is said to have lasted for seven years (two points)

23 The dry variety of Marsala is termed _____

24 Name one village which produces Vin doux naturel _____

25 The vermouths of _____ (country) are usually weathered for two years

26 A sherry warehouse is called a _____ in Spain

27 British diners when 'passing the port' traditionally hand the bottle to the person on their _____ hand side

28 Sherries which have developed the yeast 'scum' phenomena will become _____ (type of sherry)

29 Fortified wines contain approximately _____ per cent alcohol

30 A _____ measure (size) is normally used in British bars for the service of fortified wines

31 The dry coastal *fino* sherry from Sanlúcar de Barrameda is called _____

32 *Raya* sherries will develop into _____ sherries as they mature

33 Irrigation takes place in Madeira by way of skilfully designed aqueducts called _____

34 The sweetening grape for sherry is the _____ variety

35 Moscatel de Setúbal is a sweet white fortified wine from _____ (country)

36 Sherry in its first year is in the _____ stage of development

37 The French vermouth trade is centred on the city of _____

38 Sherry grapes have for years been dried out after the 'vintage' on _____ mats in the Andalucían sunshine

39 Is port fortified with brandy before fermentation, during fermentation or after fermentation? _____

40 Give the Spanish name for one type of soil found in the sherry region of Spain _____

41 Sherry in its nursery or development stage is said to be in the _____

42 Vintage port is generally considered to be ready for drinking after _____ years

43 The nutty-flavoured medium sherry which develops from a fino is called _____

44 The instrument with a handle made of whale-bone which is used for pouring sherry into a copita glass is called a _____

Points possible 50   Points obtained ____

2 *Fortified wine from many countries B*

1 Port wine is fortified after fermentation has finished   TRUE/FALSE
2 Tawny port should be decanted before service   TRUE/FALSE
3 At the table, port should traditionally be passed to the left   TRUE/FALSE
4 All port is tested and certified in Oporto before export   TRUE/FALSE
5 Régua and Pinhao are villages which produce sherry   TRUE/FALSE
6 *Fino* sherries are produced as a result of *flor* forming on the
   wine   TRUE/FALSE
7 The main sherry vineyards in Andalucía are called *quintas*   TRUE/FALSE
8 Manzanilla is a fino sherry from Sanlúcar de Barrameda   TRUE/FALSE
9 Wine from a sherry criadera is undrinkable   TRUE/FALSE
10 Vino de color is made from boiled palomino wine   TRUE/FALSE
11 Venencia is a Spanish wine-making town   TRUE/FALSE
12 Sherry casks are called 'butts'   TRUE/FALSE
13 The finest sherry is made from Muscat grapes   TRUE/FALSE
14 The driest madeira wine is called Sercial   TRUE/FALSE
15 Marsala was invented by an Englishman   TRUE/FALSE
16 Madeira is a French island   TRUE/FALSE
17 The soil of Madeira is fertile potash   TRUE/FALSE
18 Madeira wine of 1792 is said to be still drinkable   TRUE/FALSE
19 Marsala is a Spanish fortified wine   TRUE/FALSE
20 Tarragona is a sweet white fortified wine from Spain   TRUE/FALSE
21 Beaumes-de-Venise is a French fortified wine   TRUE/FALSE
22 A small bag of herbs is suspended in Commandaria St John as
   it matures   TRUE/FALSE
23 'Gin and French' refers to gin with brandy   TRUE/FALSE
24 French vermouths are usually weathered for two years   TRUE/FALSE
25 The main centre for Italian vermouth is Milan   TRUE/FALSE

Points possible 25   Points obtained _____

Section four
Eating and drinking

1 *Wine with food A*

 1 The aperitif should be a sweet wine — TRUE/FALSE
 2 Fine expensive wines should be served later in the meal — TRUE/FALSE
 3 Pasta dishes are best accompanied by German white wine — TRUE/FALSE
 4 White meats go well with white wine — TRUE/FALSE
 5 It is appropriate to recommend dry wines with dessert — TRUE/FALSE
 6 A brandy or liqueur makes a fine aperitif — TRUE/FALSE
 7 Red wine is the most suitable wine to accompany fish dishes — TRUE/FALSE
 8 Dry rosé wine would go well with salmon — TRUE/FALSE
 9 Frascati is an excellent wine to drink with shellfish — TRUE/FALSE
10 Chianti is considered suitable with beef or lamb dishes — TRUE/FALSE
11 It would be suitable to serve a *pétillant* wine with cold chicken — TRUE/FALSE
12 Game dishes need heavy, robust white wines to counter the high flavour — TRUE/FALSE
13 Red wine glasses should be filled to within $\frac{1}{4}$ inch of the top of the glass — TRUE/FALSE
14 A customer would be wise to choose a fine burgundy with curry — TRUE/FALSE
15 Sweet white Loire wines are good with fruit — TRUE/FALSE
16 Port and stilton should never be served together — TRUE/FALSE
17 Chocolate sets off excellently the flavour of wine — TRUE/FALSE
18 Light red wines go well with blue-veined cheeses — TRUE/FALSE

Points possible 18   Points obtained _____

2 *Wine with food B*

Suggest the most suitable drink to accompany the following foods. Try to avoid repeating any wine.

food   *wine*

1 Roast pheasant _____

2 Lobster salad _____

3 Savarin aux fruits _____

4 Chicken casserole _____

5 Tournedos rossini _____

6 Fried liver and bacon _____

7 Stilton cheese _____

8 Vacherin _____

9 Filet de sole Véronique _____

10 Sweetbreads of lamb _____

11 Christmas pudding _____

12 Canapés _____

13 Aunt Mary's apple pie _____

14 Steak and kidney pudding _____

15 Moules marinière _____

16 Spaghetti Napolitaine _____

17 Cream cheese _____

18 Beef curry _____

19 Cuisses des nymphes _____

20 Noisettes d'agneau fleuriste _____

21 Filet de bœuf Wellington _____

22 Tomates vinaigrette _____

23 Escalope de veau garni _____

24 Roquefort bleu _____

25 Consommé celestine _____

26 Pâté de maison _____

27 Charlotte royale _____

28 Queen of puddings _____

29 Rich fruit cake _____

30 Whitstable natives _____

As there are many wines which are correct in relation to the above foods it is not possible to give all possible answers. If your selection is similar to the example given then mark your answer correct.

Points possible 30   Points obtained _____

Section five
Bar work

1 Case studies on bar work

Charles is the new bar-keeper at the White Hart Hotel. He has been in the job for several years, having left college with qualifications in alcoholic beverages. He is a hard worker and possesses better than average social skills; as a bar-keeper, his service is fast and skilful; he is ambitious and misses no opportunity to improve his skills and expertise; he reads widely around his subject, and quickly gets to know each individual customer's likes and dislikes. During his time at the hotel there have been a number of occasions when he was called upon to think clearly and put into practice the theory he had learned at college.

Consider the following and make up your mind how Charles (and you) would react in these situations. Score four points for a correct answer. *Parts 6 and 10 of 'Table and Bar' should help you here.*

1 Mr Stevens, a local tradesman, came into the 'snug' bar and asked for a double whiskey. He was unsteady on his feet and was slurring his words. (*Part 10 of 'Table and Bar' should help here*)

2 The first couple to enter the bar in the early evening told Charles that there was a large brown paper parcel in the entrance of the hotel. What would Charles do?

3 Charles was working alone in the bar on a very busy evening. At ten minutes before closing time the draught bitter ran out. Several people were waiting to be served.

4 During a hectic service time in the public bar, a customer ordered a large round of drinks. Other people were waiting to be served. After he had been served six assorted drinks, the customer said that he had forgotten which other two he had to get, and told Charles that he would go and enquire. On the way back across the bar he was distracted and got into conversation with someone else. What would you expect Charles to do?

5 Mrs Walters complained when given her change that it was wrong. She told Charles she had given him a ten pound note and had only been given change for five pounds. How did Charles react?

6 Whilst he was collecting glasses amongst the tables in the bar Charles noticed a person obviously under 18 drinking from a pint beer mug. Charles was working alone in the bar and had not served the person in question.

7 During the early evening the telephone rang, and when Charles answered the caller asked if Mr and Mrs Roberts were in the bar. Charles knew that they were as he had just served them. What was his response?

8 After closing the bar at the end of the evening Charles found a camera on the window-sill. What would Charles have done?

9 When opening a bottle of brown ale with the counter-mounted crown cork opener Charles accidentally broke the top of the glass bottle. What happened next?

10 A friend of a resident came up to the bar whilst Charles was washing the glasses about twenty minutes after the bar had closed. He asked for a gin and tonic, a pint of lager and two packets of crisps to take into the lounge. How did Charles deal with this?

Points possible 40   Points obtained _____

2 Bar work and taking orders

1 Crown cork openers are for opening wine bottles   TRUE/FALSE
2 Spirit bottles fitted with optics are upside-down during service   TRUE/FALSE
3 The smallest jigger measure is the five-out   TRUE/FALSE
4 The three-out jigger measure is used for a double measure of spirit   TRUE/FALSE
5 It is the duty of the morning bar staff to wash the glasses from the night before   TRUE/FALSE
6 New stocks of bottles should be placed in front of the bottles already there   TRUE/FALSE
7 The spile should be placed in the shive just before service starts   TRUE/FALSE
8 Bottled guinness is pasteurised   TRUE/FALSE
9 It is illegal to give credit to drinkers for alcohol to be consumed in a bar   TRUE/FALSE
10 Eating food behind the bar is illegal   TRUE/FALSE
11 Smoking behind the bar is permitted   TRUE/FALSE
12 The duties which take place at the end of the day are called 'fermeture'   TRUE/FALSE
13 When pouring a bottle of beer, the label should face away from the purchaser   TRUE/FALSE
14 When serving lager and lime, the lime should be poured in before the beer   TRUE/FALSE
15 One bar person should be able to cope with 25 drinkers   TRUE/FALSE
16 At table, the beverage list should be presented to the host from his right   TRUE/FALSE
17 Function beverages are often chosen in advance by the organisers   TRUE/FALSE
18 Residents are usually asked to sign the check when they purchase drinks   TRUE/FALSE

Points possible 18   Points obtained _____

3 Cocktail bases

Each of the cocktails listed below are based on one of the following spirits:
(a) brandy (b) gin (c) campari (d) rum (e) vodka (f) rye whiskey (g) tequila.

Place the letter which you think is appropriate beside the name of the cocktail. If
you think a cocktail may be made with any one of a number of spirits place (h) in
the space.

1 Alexander	____	11 Pink lady	____
2 Americano	____	12 Screwdriver	____
3 Bloody Mary	____	13 Sidecar	____
4 Bronx	____	14 Sours	____
5 Champagne cocktail	____	15 Tequila sunrise	____
6 Royal clover club	____	16 Pina colada	____
7 John Collins	____	17 White lady	____
8 Cuba libre	____	18 Manhattan	____
9 Harvey wallbanger	____	19 Dry martini	____
10 Highball	____	20 Old-fashioned	____

Points possible 20   Points obtained ____

Section six
Cellar work

1 Equipment, routines and beer containers

1 Carbon dioxide gas cylinders should be strapped firmly to the wall when in use	TRUE/FALSE
2 Dip-sticks are used to prevent a cask from rolling	TRUE/FALSE
3 Filling beer casks at the brewery is called 'racking'	TRUE/FALSE
4 Spiles are placed in the shive of the cask in order to reduce or build up pressure	TRUE/FALSE
5 The tap is placed in the cask through the keystone	TRUE/FALSE
6 Casks of beer are placed vertically on the stillion or thrawl	TRUE/FALSE
7 The cellar floor should be washed with chloride of lime	TRUE/FALSE
8 Over-spill beer should be returned to the cask	TRUE/FALSE
9 The beer cellar should be kept at a constant temperature of 21°F	TRUE/FALSE
10 Strong odours in the cellar can make beer sour	TRUE/FALSE
11 Detergent in a glass will improve the head on the beer	TRUE/FALSE
12 Sunlight is harmful to beer	TRUE/FALSE
13 Sulphur tapes are burned in suspect beer casks	TRUE/FALSE
14 A nip of beer is equal to $\frac{1}{4}$ pint	TRUE/FALSE
15 Cellar beer tanks are of 90 gallons and 180 gallons	TRUE/FALSE
16 A firkin holds more than a kilderkin	TRUE/FALSE
17 A pin holds approximately 36 gallons	TRUE/FALSE
18 The tradespeople who make wooden casks are called smiths	TRUE/FALSE

Points possible 18   Points obtained ____

Section seven
Brewing and distilling

1 Brewing commodities and process

1 The best water for the production of pale ales is found at _____

2 The cereal crop which is used for the production of British beer is
_____

3 One county in which cereal for brewing is grown is _____

4 The most usual sugar used for brewing is _____ sugar

5 Mild ales in cask have sugar added to them in the form of _____
before they leave the brewery

6 Hops are members of the _____ family of plants

7 Female hop flowers contain bitter _____ which help to preserve
the beer

8 $1\frac{1}{2}$ cwt sacks called _____ are used to store hops

9 Beer is clarified by a substance obtained from fish bladders called
_____

10 Hop are dried in buildings known as _____ houses

11 Large quantities of hops are grown in the county of _____

12 Yeast is a fungus which reproduces itself by _____ (two words,
two points)

13 Bottom fermenting yeast is used in the brewing of _____

14 The brewing process starts in the _____

15 The sweet extract of malt which is fermented into beer is called
_____

16 Hops and sugar are added to the copper and the liquid will be boiled for
_____ hours

17 The fermentation of beer takes approximately _____ hours/
days/weeks (delete as appropriate, two points)

18 Used grains from the brewing process are sold as _____

19 Spent hops after boiling are sold as _____

20 The temperature after boiling is brought down in the _____
(equipment) to below _____°C (two points)

Points possible 23   Points obtained ____

2 Brewing process

1 The list below gives the eight parts of the brewing process. Try to decide in which order they occur, writing the answers in the eight boxes (representing the brewing process) provided: copper, fermenting vessel, storage tanks, wort receiver, mill, hop-back, mash-tun, paraflow.

1

2

3

4

5

6

7

8

2 Write down the three by-products that occur at stages 2, 4 and 6

(a) _____

(b) _____

(c) _____

3 List the five ingredients necessary for the manufacture of beer, and the substance used to clarify it

(a) _____   (d) _____

(b) _____   (e) _____

(c) _____   (f) _____

Points possible 17   Points obtained ___

3 Spirits

1 Alcohol vapourises at _____ °C
2 *Un trou Normand* refers to the drinking of _____
3 Calvados is made from _____ grown in _____ (area of France) (two points)
4 Do spirits improve in bottle? (Yes or no) _____
5 Which two spirits are subject to haziness in cold weather? _____ and _____ (two points) (see Part 9 of 'Table and Bar')
6 Bacardi was first made on the island of _____
7 Spirits contain approximately _____ per cent alcohol
8 _____ is a spirit made from cactus plants
9 Slivovitz is a spirit made from distilled _____
10 Vodka is purified by passing it through _____ (two words, two points)
11 Japanese _____ is traditionally served warm in porcelain cups
12 Irish potato spirit distilled illegally is known as _____
13 Aniseed-flavoured spirit which turns milky when mixed with water is called _____ (France) or _____ (Greece) (two points)
14 Arrack is a popular spirit made from the sap of _____
15 Absinthe is a harmful spirit illegally distilled from _____ plants
16 Fernet branca is a bitters. It is said to be particularly good for curing _____ and _____ (two points)
17 Kirsch is a spirit from the _____ (two words, two points) area of Germany
18 Scotch whisky owes its predominance over brandy on world markets to the disease _____
19 Slaves were responsible for making the first _____
20 German imitation rum is known by the name rum _____
21 Grande champagne is a fine quality _____
22 Sylvius Van Leyden invented _____
23 Jim Beam is a type of _____ whiskey
24 'Mother's ruin' was a name used for _____
25 Gin was first made in 1577 in _____ (country)
26 Irish whiskey is distilled in a pot still _____ times (number)
27 Name a botanical used to flavour gin _____
28 Bourbon whiskey is made from a mixed mash of cereals of which 51 per cent must be _____
29 Grog was a mixture of _____ and water
30 The patent still was invented by _____

31 Highland malt whisky distilleries are found north of an imaginary line between _____ and _____ (towns) (two points)

32 The finest Highland malt whiskies are reputed to come from an area along the river _____

33 The best calvados comes from the region of _____

34 A puncheon is a cask holding 110 gallons of _____

35 The result of the first distillation of cognac is called _____

36 Sugar-cane is the commodity used in the manufacture of _____

37 The colour of dark rum is caused by adding _____

38 Name a colourless rum _____

39 The predominant cereal in Canadian club whiskey is _____

40 Grain whisky is distilled in a _____ still

41 Blending of malt whisky with grain whisky began in the year _____

42 _____ (place name) whisky distilleries are situated at the end of the Mull of Kintyre

43 Name a German brandy _____

44 Cognac brandy is matured in casks made of _____ oak

45 The first and the last parts of the pot-still distillation are known as _____ and _____ (two points)

46 Grappa is a cheap brandy made in _____

47 Schnapps is distilled in Germany and Holland from fermented _____

Points possible 55   Points obtained ____

4 *Raw materials of spirits*

Write in the space provided the predominant raw material from which the following spirits are made.

1 Tequila	_____	14 Canadian Club	_____
2 Quetsch	_____	15 Schnapps	_____
3 Arrack	_____	16 Framboise	_____
4 Calvados	_____	17 Mirabelle	_____
5 Poteen	_____	18 Aquavit	_____
6 Kirsch	_____	19 Bacardi	_____
7 Vodka	_____	20 Armagnac	_____
8 Scotch	_____	21 Bourbon	_____
9 Grappa	_____	22 Asbach	_____
10 Poire William	_____	23 Irish whiskey	_____
11 Cognac	_____	24 Dimple Haig	_____
12 Sake	_____	25 Tropicana	_____
13 Daiquiri	_____	26 Slivovitz	_____

Points possible 26   Points obtained ____

5 *Beer, cider and spirits*

1 During germination of barley, insoluble starch is converted to soluble sugar   TRUE/FALSE
2 Norfolk is noted for the production of large quantities of high quality hops   TRUE/FALSE
3 It is the male flower of the hop plant which contains the bitter dust *lupalin*   TRUE/FALSE
4 Hops are usually stored in pockets   TRUE/FALSE
5 When more hops are used in a brew the beer will last longer   TRUE/FALSE
6 Spent barley grains are sold as fertilizer   TRUE/FALSE
7 A paraflow is used to pump beer from the cellar to the bar   TRUE/FALSE
8 Pulp from apples which is pressed for cider is called the 'cheese'   TRUE/FALSE
9 Perry is made from a special kind of apple   TRUE/FALSE
10 A spirit safe is where very old bottles of malt whisky are kept   TRUE/FALSE
11 The patent still was invented in Dublin   TRUE/FALSE
12 Spirits can be made from any commodity which has fermented with yeast   TRUE/FALSE
13 Water vapourises at a lower temperature than alcohol   TRUE/FALSE
14 Methyl alcohol is safe to use in cocktails   TRUE/FALSE
15 Gin was first made in Switzerland   TRUE/FALSE
16 Prohibition was a law which was passed to stop alcohol abuse in London   TRUE/FALSE
17 Plymouth gin is a correct ingredient of a 'pink gin'   TRUE/FALSE
18 Gin is flavoured with juniper berries   TRUE/FALSE
19 Spirits at drinking strength usually contain 70 per cent alcohol   TRUE/FALSE
20 Martell, Hine and Hennessy were all from the British Isles   TRUE/FALSE
21 St Emilion is a grape variety used for cognac manufacture   TRUE/FALSE
22 *Bagasse* is the name used for crushed sugar cane in the West Indies   TRUE/FALSE
23 Captain Morgan is a rum from Trinidad   TRUE/FALSE
24 *Phylloxera* gave Scotch an advantage over cognac   TRUE/FALSE
25 Bourbon whiskey is matured with charred oak   TRUE/FALSE

Points possible 25   Points obtained ____

6 *Speciality coffees*

Write in the space provided the name of the spirit or liqueur which would correctly be used in the following speciality coffees. Award one point for each correct answer.

1 Irish   _____

2 Caribbean   _____

3 Calypso   _____

4 Balalaika   _____

5 Highland   _____

6 Monk's   _____

7 Café Napoleon   _____

8 Café royale   _____

9 Witch's   _____

10 Yorkshire   _____

11 Gaucho's   _____

12 Kentucky   _____

13 Bonnie Prince Charlie   _____

Points possible 13   Points obtained _____

7 *Classification of liqueurs*

Below are listed 25 liqueurs which can each be classified as one of the following types: (a) herb (b) fruit (c) coffee (d) chocolate (e) nut and kernal (f) whisky and cream (g) egg (h) liqueur containing twigs or flowers.

Place the letter you think appropriate next to the name of the liqueur. Score one point for each correct answer.

1 Kahlua	____	19 Fraisia	____
2 Yellow Chartreuse	____	20 Bahia	____
3 Van der Hum	____	21 Amaretto di Saronno	____
4 Noyau	____	22 Carolans	____
5 Advocaat	____	23 Sapin d'Or	____
6 Kümmel	____	24 Senancole	____
7 Vieille cure	____	25 Bénédictine	____
8 Cointreau	____	26 Grand Marnier	____
9 Cassis	____	27 Cordial Médoc	____
10 Palo	____	28 Glayva	____
11 Merlyn	____	29 Fior d'Alpi	____
12 Malibu	____	30 Crème de Cacao	____
13 Tia Maria	____	31 Parfait Amour	____
14 Edelweiss	____	32 Southern Comfort	____
15 Midori	____	33 Glen mist	____
16 Galliano	____	34 Green Chartreuse	____
17 Brontë	____	35 Izzara	____
18 Curaçao	____	36 Crème de Menthe	____

Points possible 36   Points obtained ____

8 Cocktails and liqueurs

You may need to refer to parts 8 and 6 of 'Table and Bar' here.

1 Cocktails should be shaken when the ingredients are _____

2 If all the ingredients are clear, then the cocktail should be _____

3 An 'Americano' cocktail contains Campari and _____

4 The spirit ingredient of a 'Bronx' cocktail is _____

5 A 'Cuba libre' cocktail is made up of rum, coca cola and _____

6 A 'Harvey wallbanger' cocktail is completed by gently pouring _____ over the drink

7 A 'Pina colada' consists of white rum, coconut cream and _____

8 The liqueur which is used in a 'White lady' cocktail is _____

9 _____ is used to give pink gin its pink colour

10 Tequila, orange juice and grenadine are the ingredients of _____

11 The spirit used in an 'Egg-sour' is traditionally _____

12 The liquid ingredients of a 'Screwdriver' cocktail are _____ and _____ (two points)

13 A 'Sidecar' and a 'White lady' are made of the same ingredients except for the spirit. In a 'White lady' it is _____ while in a 'Sidecar' the spirit is _____ (two points)

14 'Kir' is a blend of dry white burgundy wine and _____

15 The best known 'highball' cocktail is the _____

16 The spoon which is used in the bar for stirring cocktails is called a _____

17 Brandy, crème de cacao, and fresh cream are the ingredients of an _____ cocktail

18 The liqueur with the highest alcohol content is _____

19 Kümmel is flavoured with _____

20 A Yorkshire liqueur in an earthenware jar is called _____

21 Which liqueur is said to have legendary power to keep a couple from parting if they drink it together? _____

22 Write down the colour of cointreau _____

23 Name a liqueur made from strawberries _____

24 Name the liqueur made from brandy, egg-yolks, and sugar _____

25 Midori is made from _____

26 Parfait amour is _____ or _____ in colour (two points)

27 Kahlúa is made from _____ and _____ (two points)

28 The liqueur which has a connection with Bonnie Prince Charlie is
_____

29 Name a liqueur containing sugared twigs _____

30 Calypso coffee includes the liqueur _____

9 *Temperatures and strengths*

Each of the following pairs represents an important temperature or strength connected with alcohol. (*Information about temperatures will be found in Part 3 of 'Table and Bar'.*)

Temperatures (a) 22 °F (–6 °C) (b) 50 °F (10 °C) (c) 65 °F (17 °C) (d) 172 °F (78 °C)
Strengths (e) 10 °OIML (17 °Sikes) (f) 20 °OIML (35 °Sikes) (g) 30 °OIML (53 °Sikes) (h) 40 °OIML (70 °Sikes) (i) 70 °OIML (123 °Sikes) (j) 100 °OIML (175.1 °Sikes)

Match these temperatures or strengths to the definitions given below. Score *two* points for each correct answer.

1 Evaporation of alcohol during distillation   _____

2 Formation of crystals of tartar   _____

3 Maturing strength of pot-still spirits   _____

4 Usual drinking strength of spirits   _____

5 Strength of pure alcohol   _____

6 Serving temperature for white and rosé wines   _____

7 Serving temperature for red wine   _____

8 Approximate strength of table wine   _____

9 Approximate strength of pot-still spirits after first
distillation   _____

10 Approximate strength of fortified wine   _____

Points possible 20   Points obtained _____

Section eight
Storage of alcohol and other items

1 Storage, control and tobacco

1 Light ale has a longer life expectancy than _____ ale
2 Spirits are best stored _____
3 Liqueurs should be kept out of _____ (two words, two points)
4 White wines should be stored nearest to the _____
5 Bottles of vintage port should be stored on their sides with the _____ uppermost (two words, two points)
6 Vintage champagne may keep for up to _____ years
7 Vintage red wines will keep from _____ to _____ years (two points)
8 Moselle, Alsace and Beaujolais wines are best drunk between _____ and _____ years (two points)
9 Wine should be stored at a constant temperature of between _____° and _____° (two points)
10 The correct storage temperature for tobacco is _____°
11 An _____ stock-taking is required by law in licensed premises
12 The _____ function on an electronic till is for reading and resetting the machine and clearing it
13 In 1492 _____ sent men ashore in North America where they found people smoking rolled-up leaves
14 Briar pipes are made from _____ (plant)
15 Cigarettes were invented in _____ (country)
16 It is generally accepted that the best quality cigar tobacco is grown in _____ (country)
17 Black tobacco which is used in cocktail cigarettes is called _____ (two words, two points)
18 _____ wood is used to manufacture cigar boxes
19 Imported tobacco must remain in a bonded warehouse for _____ years
20 The cigar band should be removed by the waiter by _____

Points possible 26   Points obtained ____

Section nine
The law, health and safety

1 Law, fire, pests, cleaning, health and safety

1 A gill is $\frac{1}{5}$ of a pint   TRUE/FALSE
2 It is illegal to drive with more than 80 mg of alcohol in
   100 ml of blood   TRUE/FALSE
3 Children under 14 must *never* enter a public bar   TRUE/FALSE
4 The passing of betting slips in a public bar is permitted   TRUE/FALSE
5 An optic is the measure on a wine or spirit bottle dispensing a
   predetermined quantity   TRUE/FALSE
6 Staff should receive verbal fire-fighting instructions every six
   months   TRUE/FALSE
7 Modern fire blankets are made of fibre glass   TRUE/FALSE
8 Water jet extinguishers are suitable for fat fires   TRUE/FALSE
9 Carbon dioxide extinguishers are red   TRUE/FALSE
10 Dry powder extinguishers are blue   TRUE/FALSE
11 Water sprinkler systems are plumbed into floors   TRUE/FALSE
12 Moths are attracted into premises by bright lights   TRUE/FALSE
13 Electric ultra-violet treatment is commonly used for cockroach
   infestation   TRUE/FALSE
14 Spiders prefer warm, dry rooms   TRUE/FALSE
15 Bar-swabs should soak overnight in disinfectant   TRUE/FALSE
16 Food in the stomach speeds up the absorption of alcohol into
   the blood   TRUE/FALSE
17 Alcohol is extremely harmful in the first four months of
   pregnancy   TRUE/FALSE
18 Between 2200 and 0400 hours more than $\frac{2}{3}$ of drivers involved
   in accidents are over the legal limit   TRUE/FALSE
19 Knees should be straight when lifting heavy objects   TRUE/FALSE
20 A record must be kept of every accident, however slight   TRUE/FALSE

Points possible 20   *Points obtained* _____

Section ten

General section (Research and further reading will be required for a few of the questions in this section).

1 Grape varieties

Consider these lists of the more famous wine-making grape varieties:

black grapes
- (i) Nebbiolo
- (ii) Syrah
- (iii) Spätburgunder
- (iv) Gamay
- (v) Pinot Noir
- (vi) Zinfandel
- (vii) Cabernet Sauvignon
- (viii) Grenache
- (ix) Merlot
- (x) Sangiovese
- (xi) Pinot Meunier
- (xii) Cabernet Franc
- (xiii) Kadarka

white grapes
- (a) Sémillon
- (b) Sauvignon
- (c) Riesling
- (d) Folle Blanche
- (e) Gewurztträminer
- (f) Palomino
- (g) Chardonnay
- (h) Chenin Blanc
- (j) Aligoté
- (k) Malvasia
- (l) Pedro Ximénez (Pedro X)
- (m) Sylvaner
- (n) Viognier
- (o) Muscat

Listed below are 14 wine regions of the world, and opposite to each are lines indicating the number of grape varieties (from the above lists) commonly grown in that district.

Write the appropriate numbers or letters from the lists on the lines opposite the region. Score one point for each correct answer.

1 Champagne ____ ____ ____

2 Sherry region ____ ____

3 Bordeaux (black) ____ ____ ____

4 Bordeaux (white) ____ ____

5 Alsace ____ ____ ____

6 Burgundy ____ ____ ____

7 Italy ____ ____ ____

8 Germany ____ ____ ____

9 Rhône ____ ____ ____

10 California ____ ____ ____

11 Loire ____ ____ ____

12 Cognac ____

13 Madeira ____

14 Beaujolais ____

Points possible 34   Points obtained ____

2 *Capacities and measures*

Each of the items numbered below is incorrectly linked with a measure.

Write the correct measure in the space provided. Answers should be chosen from the list in the right-hand column. Score one point for each correct answer.

1	Shipping butt of sherry	70 cl	_____
2	Pipe of port	20 bottles	_____
3	Bottle of fortified wine	100 cl	_____
4	Bottle of gin	18 gallons	_____
5	Bottle of Tokay Aszu	9 gallons	_____
6	Tank of beer (large)	4 bottles	_____
7	Magnum	115 gallons	_____
8	Kilderkin	108 gallons	_____
9	Bottle of beaujolais	36 gallons	_____
10	Tank of beer (small)	16 servings	_____
11	Jeroboam	2 bottles	_____
12	Barrel	32 servings	_____
13	Litre of wine	54 gallons	_____
14	Firkin	50 cl	_____
15	Nebuchadnezzar	180 gallons	_____
16	Hogshead	90 gallons	_____

Points possible 16   Points obtained _____

3 *Geographical locations*

Each of the items numbered below (and over the page) is incorrectly linked with the geographical location in the centre column. The locations listed in the centre column refer to one of the commodities in the left-hand column.

Write the correct location in the space provided. Score one point for each correct answer.

1 Champagne	Vosges Mountains	_____
2 Sherry	River Ebro	_____
3 Port	Australia	_____
4 Cognac	Funchal	_____
5 Châteauneuf-du-Pape	Speyside	_____
6 Pommard	Bahamas	_____
7 Claret	Saar and Ruwer	_____
8 Gewurztträminer	Grande Champagne	_____
9 Calvados	Savoie	_____
10 Le Chablais	Franconia	_____
11 Pecs rosé	Amsterdam	_____
12 Mosel	South Africa	_____
13 Schloss vollrads	Veneto	_____
14 Rioja	Hampshire	_____
15 Chianti	Troodos mountains	_____
16 Soave	Jura	_____
17 Lacryma Christi	Zimbabwe	_____
18 Hambledon	Midi	_____
19 Zinfandel	Yorkshire	_____
20 KWV paarl	Trinidad	_____
21 Coonawarra claret	River Rhine	_____
22 Sercial	Tuscany	_____
23 Commandaria	Normandy	_____
24 Gin	River Gironde	_____
25 Scotch	Latium	_____
26 Martini	Vaud	_____
27 Bacardi	River Minho	_____
28 Angostura bitters	River Ill	_____
29 Lutomer Riesling	Touraine	_____
30 Vinho verde	California	_____
31 Frascati	Fleurie	_____
32 Steinwein	Yugoslavia	_____
33 Burgundy	Andalucía	_____

34 Chambéry	Campania	_____
35 Vin jaune	Rheingau	_____
36 St Raphael	Hungary	_____
37 Liebfraumilch	Turin	_____
38 Chinon	Côte de Beaune	_____
39 Beaujolais	Côte des Blancs	_____
40 Cigarette tobacco	River Saône	_____
41 Alsace	Provence	_____
42 Minervois	Upper Douro	_____
43 Brontë	Rhône valley	_____

Points possible 43   Points obtained _____

4 *Alphabet quiz*

Each solution to the clues begins with a different letter of the alphabet.

1 A _____ Chalky soil in the sherry region
2 B _____ Mouldy condition of the grapes for Sauternes
3 C _____ Main flavouring ingredient of Kümmel
4 D _____ Sweetness in Spanish wine
5 E _____ Wood used to make briar pipes
6 F _____ Glass for champagne
7 G _____ Dutch gin
8 H _____ A beer cask
9 I _____ Island producing malt whisky
10 J _____ Botanical used to flavour gin
11 K _____ Grower's choice of German wine
12 L _____ Irrigation in Madeira
13 M _____ Treaty with Portugal
14 N _____ River region producing German wine
15 O _____ Spirit made in Greece
16 P _____ Disease of the vine
17 Q _____ Vineyard in the Upper Douro region of Portugal
18 R _____ Daily shaking of champagne bottles
19 S _____ Tokay made with non-noble-rot grapes
20 T _____ Acid crystals which may form in wine
21 U _____ A type of bitters from Germany
22 V _____ White wine from Touraine in the Loire valley
23 W _____ Shrub commonly used to aromatise vermouths
24 X _____ Sweetening grape for sherry (after Pedro)
25 Y _____ Essential for fermentation to take place
26 Z _____ Red premium wine from California

Points possible 26   Points obtained _____

5 *Initials quiz*

In the spaces provided, write down what the initials stand for, and add a short definition or explanation. Score one point for each correct answer.

1 VSOP   (a) _____
            (b)

2 KWV   (a) _____
            (b)

3 AOC   (a) _____
            (b)

4 QmP   (a) _____
            (b)

5 OIML   (a) _____
            (b)

6 VDQS   (a) _____
            (b)

7 DOC   (a) _____
            (b)

8 CO_2 (a) _____
            (b)

9 LBV   (a) _____
            (b)

10 VDN   (a) _____
            (b)

11 UKBG   (a) _____
            (b)

12 CAMRA   (a) _____
            (b)

13 NV   (a) _____
            (b)

14 LVA   (a) _____
            (b)

15 HCIMA   (a) _____
            (b)

16 CB   (a) _____
            (b)

17 QbA   (a) _____
            (b)

Points possible 34   Points obtained ____

6 *Initials and clues*

The following groups of letters represent short words or phrases connected with alcoholic beverages. In each case you are given a clue and the number of the page in *Table and Bar* where the information may be found.

1 LVA. Formed to assist the licencee (3) _____

2 HCIMA. Professional association (3) _____

3 FH. Privately owned public house (4) _____

4 VV. European grape species (10) _____

5 CS. Used as a spray to prevent mildew (10) _____

6 FE. Removes grape stalks (11) _____

7 DP. Inventor of champagne (13) _____

8 MDR. Area where champagne grapes grow (15) _____

9 MC. Method of making superior sparkling wine (15) _____

10 EOA. Married Henry, Duke of Dijon (18) _____

11 CS. Black grape of Bordeaux (19) _____

12 BOC. When the English lost Bordeaux (19) _____

13 FR. Unrest in Burgundy and Paris in 1789–1791 (20) _____

14 BN. Wine in annual race in November (21) _____

15 HDB. Supported by an auction (21) _____

16 LAT. Grown with vines at Châteauneuf-du-Pape (22) _____

17 PSL. Dry white wine from central Loire (26) _____

18 LP. Scientist from the Jura (31) _____

19 VJ. French table wine which develops *flor* (31) _____

20 QmP. German quality wine (32) _____

21 EEE. Italian white wine (35) _____

22 JVR. Planted first vines in South Africa (42) _____

23 LK. South African wine region (43) _____

24 JB. Developed Australian vineyards (44) _____

25 GR. Led to planting of vines at Rutherglen (44) _____

26 EVA. Give assistance to British growers (51) _____

27 TC. Famous Gloucestershire vineyard (51) _____

28 DG. Cause of recent wine scandal (51) _____

29 GW. Swiss wines matured in caves (52) _____

30 MEP. Everything ready for service (67) _____

31 VNDG. Where port wine is stored (76) _____

32 UD. Mountains of north Portugal (77) _____

33 SS. Fractional blending system for sherry (79) _____

34 ES. Used in Madeira for heating wine (82) _____

35 PDC. Fortified wine from Cognac area (84) _____

36 EDM. Dry white wine from Bordeaux (88) _____

37 HFM. Robust red wine is needed to accompany (89) _____

38 WWS. A non-pasteurised beer (96) _____

39 BF. May have made the first cocktails (100) _____

40 AC. Notorious during prohibition (101) _____

41 MS. Used to stir cocktails (105) _____

42 DS. Used to measure the beer left in the cask (109) _____

43 COL. Wash the cellar floor with it (112) _____

44 OH. Buildings where hops are dried (120) _____

45 SC. Bottom-fermenting yeast (121) _____

46 WR. Immediately after the hop-back (122) _____

47 SP. Natural mineral water from Italy (125) _____

48 JP. He made the first artificial mineral water (125) _____

49 SS. Where heads and tails are separated (126) _____

50 AC. Invented the patent still (127) _____

51 CAEO. Collects the duty (127) _____

52 HR. Very pure spirit (128) _____

53 SVL. Made the first gin in Amsterdam (129) _____

54 CCG. Cheap gin (130) _____

55 GC. Best growing area for cognac vines (131) _____

56 LO. Special wood for cognac casks (132) _____

57 RV. Imitation rum (133) _____

58 WD. Clear rum (134) _____

59 URM. Dundee maturers of rum (134) _____

60 AV. Known as 'Old Grog' (135) _____

61 JD. Popular bourbon brand (137) _____

62 OBM. Well-known Irish whiskey (137) _____

63 UTN. Calvados for drinking during the meal (138) _____

64 AC. Used to refine vodka (140) _____

65 CR. Coffee with cognac and cream (142) _____

66 GC. Strongest liqueur (144) _____

67 VC. One of the oldest herb liqueurs (144) _____

68 FF. Liqueur made from grapefruit (145) _____

69 SC. Liqueur based on bourbon whiskey (146) _____

70 BIC. Liqueur from Ireland containing cream (147) _____

71 PA. Measured with a refractometer (148) _____

72 UP. Below 100° on the Sikes scale (148) _____

73 CH. Condition of beer in very cold weather (152) _____

74 CN. Given by supplier for returns and empties (154) _____

75 BC. Used for accurate stock recording (154) _____

76 CC. His sailors found men smoking (156) _____

77 GP. Time when boys where whipped for not smoking (157) _____

78 BW. Where tobacco must remain for two years (158) _____

79 CWB. Storage containers for cigars (158) _____

80 PP. Your appearance and attitude (160) _____

81 TW. Necessary in a group of workers (161) _____

82 SDA. Ensures that females are treated equally (168) _____

83 BAL. 80 mg of alcohol in 100 ml of blood (168) _____

84 FB. Used to exclude air from fat-pan fires (171) _____

85 DPE. Are coloured blue and may be used for most fires (173)_____

86 BA. Use 'nippon' to destroy them (175) _____

87 CV. Useful when you apply for a position (188) _____

7 *Companies*

Below you will see the names of 26 companies who are each associated with a particular well-known product.

Identify the item for which they are most famous, writing the answer in the space provided.

1 Moët & Chandon _____

2 Stones _____

3 Gordons _____

4 Lambs _____

5 Jim Beam _____

6 Heering _____

7 Hennessy _____

8 Schweppes _____

9 Bulmers _____

10 Whitbread _____

11 Harveys _____

12 Thwaites _____

13 Britvic _____

14 Warnincks _____

15 Haig _____

16 Gonzalez Byass _____

17 Freemans _____

18 Tuborg _____

19 Warre _____

20 Watneys _____

21 Jamesons _____

22 Roses _____

23 Hine _____

24 Schmirnoff _____

25 Booths _____

26 Lanson _____

Points possible 26   Points obtained _____

8 *Speed test*

Write down the meaning of the following in no more than *four* words and in less than four minutes. (The page number in *Table and Bar* is shown in brackets.)

1 Angelica (83)
2 *Bodega* (79)
3 Commandaria St John (83)
4 *Dosage* (16)
5 *Edelfäule* (32)
6 Fins bois (131)
7 Grist (121)
8 Hautvillers (13)
9 Isinglas (121)
10 Jeroboam (17)
11 Kir (104)
12 Loupiac (19)
13 Manzanilla (78)
14 *Natur* (16)
15 Old crow (137)
16 Prohibition (130)
17 Quetsch (140)
18 Rainwater (82)
19 Swizzle-stick (14)
20 Tinajars (38)
21 Umbria (35)
22 Vermouth (83)
23 Woodhouse (84)
24 XO (132)
25 Zwicker (26)

Points possible 25   Points obtained ____

9 *Crosswords*

Crossword 1

Across

1 Disastrous sudden frosts may occur (5)
5 Modern casks (5)
8 Attempted to recognise the bouquet (5)
9 Government stamped beer glasses (5)
10 From the mill (5)
11 Wines from Sancerre (9)
13 Southwards through Burgundy (5)
15 Measures potential alcohol (13)
16 Shelves in the cellar where white wines may be stored (5)
19 Bouquet (5)
22 Keep the black grapes separate from the white ones (5)
23 Fruit of Beaujolais (5)
24 Illegal distiller _ _ _ _ _ his still when authorities call (5)
25 Levels of distinction (5)
26 Stride over the ground again (5)

Down

1 Start afresh (2–3) (two points)
2 Report on Cyprus wine industry (5)
3 Where casks of port are stored (5)
4 Postprandial beverage (7–6) (two points)
5 Pale, crystal or black (5)
6 Spanish red (5)
7 Port warehouse (5)
12 French earth (5)
14 Correct (5)
16 Bottom fermented (5)
17 Protected by the Sex Discrimination Act (5)
18 Basic sherries (5)
19 Not this one (5)
20 These wines may contain sediment (5)
21 From Greek pine-forests (5)

Points possible 32   Points obtained _____

Crossword 2

Across

1 22 January (2–8–3) (three points)
8 Roadside stopping place (3)
9 Most German wine (in German) (9)
10 Beer drinker's throat (gutter)? (3–5) (two points)
11 Not fuissé (4)
14 Worth a drink in the (golf) club-house (1–3–5) (3 points)
17 Vineyard workers rushed at vintage time (4)
18 Pleads with (8)
20 Flavoured spirit (6–3) (2 points)
22 Best export light ale – initials (3)
23 From individual grapes (13)

Down

1 Twist around the binds (5)
2 Country wine (3–2–4) (three points)
3 Fresh plantings of lesser known wines (3–5) (two points)
4 Home of lambrusco (6–7) (two points)
5 Scientific evaluation of wine (5)
6 Delivery expected (3)
7 Over there (6)
12 These grapes usually produce *pétillant* wine (5–4) (two points)
13 Clear empty ones, but leave these glasses on the table (4–4) (two points)
15 Blood will _ _ _ _ _ alcohol (5)
16 Bottled where it was grown (6)
19 Some people's impression of the (monstrous?) sommelier (4)
21 Hopped beer (3)

Points possible 36   Points obtained ____

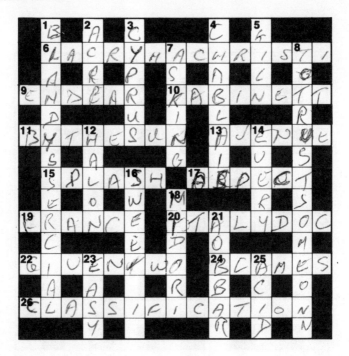

Crossword 3

Across

6 Wept for Naples (7–7) (two points)
9 To cause to be beloved (6)
10 Grower's own choice of wine (8)
11 Ripened ... (2–3–3) (three points)
13 A meeting place with trees (6)
15 Served with spirit (6)
17 Towards the sun (6)
19 Renowned for fine wines (6)
20 Largest producer and its label language (5–3) (two words) (two points)
22 A couple of free pints bought for you (5–3) (two points)
24 Allocates guilt (6)
26 1855 tables (14)

Down

1 Popular Madeira (7–7) (two points)
2 Area of land no longer used (4)
3 Land of Aphrodite (6)
4 Swiss, definitely not Burgundy (8)
5 Place for drying-up (4)
7 Making requests (6)
8 To have complete faith in the bar-person (or anyone else) (2–5–7) (three points)
12 Green extinguisher (5)
14 Each one without exception (5)
16 Old Tom (5–3) (two points)
18 Green product from the land of the rising sun (6)
21 Any other business before cocktails (3–3) (two points)
23 No difficulty (5)
25 Tart taste in wine (4)

Section eleven
Multiple-choice questions

Part 1 A service industry

1 The hotel, catering and leisure industries employ, in the UK
a less than $\frac{1}{2}$ million people
b between $\frac{1}{2}$ million and $1\frac{1}{2}$ million
c between $1\frac{1}{2}$ million and 2 million
d over 2 million people
2 A free house is a public house or licensed hotel where
a residents are accommodated free of charge
b the manager can buy his beers, wines and spirits from any source
c the establishment is granted its licence free of charge
d normal licensing regulations do not apply
3 A tenant of a public house
a pays rent   c is paid a salary
b owns the licence   d works for the licencee
4 The most popular alcoholic beverages are
a wine in bars and spirits in restaurants
b wine in restaurants and draught beer in bars
c spirits in bars and bottled beer in restaurants
d liqueurs in restaurants and bottled beers in bars

Part 2 Wine from many countries

1 Fortified wines are those to which has been added
a brandy b vitamins c sulphur d sugar
2 Which of the following countries is situated within the wine-making latitudes?
a New Zealand b India c Guyana d Mexico
3 Vines will produce better quality grapes for wine-making if they are on slopes facing
a north b east c south d west
4 Black grapes grow on vines of the variety
a Cabernet Sauvignon b Sylvaner c Riesling d Chardonnay
5 *Phylloxera* was first discovered in 1858 in
a England b France c California d Germany
6 Sweet wines will be produced if the vines are affected by
a oidium b downy mildew c *botrytis cinerea* d black-rot
7 A *foulloir-egrappoire* is used to
a remove stalks   c clarify wine
b pick grapes   d separate wine from sediment
8 The French wine-making region which is close to the German border is
a Rhône b Loire c Alsace d Provence

9 Central Loire is famous for the production of
 a fine dry white wines   c quality red wines
 b medium sweet rosés   d VDQS red wines

10 The region of North Burgundy is often called
 a Côte Rôtie b Côte d'Or c Côte des Blancs d Côte Mâconnaise

11 The soil of Champagne is mainly
 a slate b chalk c gravel d clay

12 Dom Pérignon experimented on clarification and sparkle in the abbey at
 a Bouzy b Ay c Hautvillers d Reims

13 The blending together of the village wines of Champagne is called
 a *rémuage* b *tirage* c *dosage* d *assemblage*

14 The number of casks which may by law be produced from one pressing (marc) of champagne grapes is
 a 10 b 11 c 12 d 13

15 *Natur* on the label of a bottle of champagne indicates that the wine is
 a sweet b medium sweet c dry d very dry

16 A bottle of champagne will be very sweet if the label contains the word
 a doux b sec c brut d demi-sec

17 A large bottle of champagne containing the equivalent of four bottles of wine is called a
 a methuselah b jeroboam c magnum d balthazar

18 Charmat invented a method of secondary fermentation which takes place in
 a bottles b barrels c sealed tanks d open vats

19 The ideal temperature for the service of champagne is
 a 45 °F (7 °C) b 65 °F (17 °C) c 50 °F (10 °C) d 55 °F (13 °C)

20 The Bordeaux vineyard area belonged to England for 299 years because of a
 a treaty b marriage c war d government decree

21 One of the best red wine areas of the world is
 a Sauternes b Entre-deux-Mers c Loupiac d Médoc

22 English forces were defeated and lost Bordeaux at the battle of
 a Bordeaux b Monbazillac c Castillon d Pomerol

23 Which of the following is a *premiers grands crus classés* wine of the 1855 Médoc classification?
 a Château Lafite-Rothschild   c Château Ausone
 b Château Palmer   d Château Montrose

24 Château Mouton-Rothschild was promoted to the *premiers grands crus classés* in
 a 1901 b 1928 c 1969 d 1973

25 Burgundy vineyards tend to be a small because of
 a church influence   c The French Revolution
 b Napoleon's battles   d steep terracing

26 Passe-tout-grains is a wine blend of
 a one-third Chardonnay and two-thirds Aligoté
 b one-third Pinot noir and two-thirds Gamay
 c two-thirds Chardonnay and one-third Aligoté
 d two thirds Pinot noir and one-third Gamay

27 The Hospices de Beaune auction is held each year in November on the
 a first Sunday   c first Thursday
 b third Sunday   d third Thursday

28 The wines of north Burgundy are made from the grape variety
 a Pinot Noir b Cabernet Sauvignon c Gamay d Merlot

29 Which of the following villages of Burgundy is furthest north?
   a Morgon   b Givry   c Gevrey-Chambertin   d Meursault
30 Which of the following villages is furthest south?
   a Juliénas   b Pommard   c Volnay   d Aloxe-Corton
31 The northern Rhône valley is
   a very flat   b undulating   c steeply terraced   d a flood plain
32 Sparkling wines made by the *méthode champenoise* are produced around the
   village of
   a Hermitage   b St Péray   c St Joseph   d Cornas
33 Which of the following wines is a VDN wine of the Rhône valley?
   a Châteauneuf-du-Pape   b Tavel   c Beaumes-de-Venise   d Gigondas
34 Lavender and thyme are grown by tradition in the vineyards of
   a Châteauneuf-du-Pape   b Lirac   c Crozes Hermitage   d Cornas
35 Châteauneuf-du-Pape must never be sold with an alcoholic strength of less
   than
   a 10 per cent   b 11 per cent   c 12.5 per cent   d 13 per cent
36 The vineyards of Alsace are near to the river
   a Ill   b Marne   c Chârente   d Dordogne
37 Which of the following is a vine variety commonly grown in Alsace?
   a Chardonnay   b Sylvaner   c Aligoté   d Semillon
38 *Zwicker* indicates that the wine is
   a for export only   c suitable for diabetics
   b very old   d from a blend of grapes
39 Most Alsace wine should be drunk
   a in the year of production
   b while it is less than eight years old
   c after 10 years maturing in wood
   d after 20 years in bottle
40 From the central vineyards of the Loire to the coast is approximately
   a 100 miles   b 300 miles   c 600 miles   d 800 miles
41 Which of the following towns is famous for its riding academy as well as its
   sparkling wine?
   a Saumur   b Chinon   c Angers   d Orléans
42 The light red wine of the Loire which is sometimes served chilled is
   a Beaujolais   b Vouvray   c Sancerre   d Chinon
43 Which wine, from this selection would go best with shellfish?
   a Bourgueil   b Montlouis   c Muscadet   d Coteaux du Layon
44 Which Loire town produces excellent wine vinegar?
   a Orléans   b Tours   c Angers   d Nantes
45 Select the dry white wine which is made in central Loire
   a Pouilly-Fuissé   b Pouilly-Fumé   c Pouilly-Loche   d Pouilly-Vinzelles
46 The largest in area of all the VDQS *appelations* of the southern region is
   a Languedoc   b Minervois   c Corbières   d Provence
47 Wines of Fitou are usually
   a dry white   b sweet white   c red   d rosé
48 The most common grape varieties in the Midi region are the
   a Syrah and Viognier   c Grenache and Carignan
   b Cabernet Sauvignon and Merlot   d Semillon and Sauvignon
49 The best known wine of Savoie is
   a Crépy   b Grenoble   c Seyssel   d Chambery
50 Chambery is famous for its

a red table wine   b white table wine   c vermouth   d sparkling wine
51 The famous rosé wine of the Jura is
   a vin gris   b vin fou   c vin jaune   d vin de paille
52 The Jura wine which develops a *flor* yeast scum on its surface during fermentation is
   a vin gris   b vin fou   c vin de paille   d vin jaune
53 The Jura is famous as the home of
   a Louis Pasteur   c Baron Le Roy de Boiseaumarie
   b Charles de Gaulle   d Dom Pérignon
54 Which of the following wines are traditionally served in green-stemmed glasses?
   a Hock and Steinwein   c Mosel and Alsace
   b Ahr and Rheingau   d Steinwein and Nahe
55 Palatinate is another name for
   a Rheingau   b Rheinpfalz   c Mittel-Rhein   d Rheinhessen
56 The famous German wine which originated from the city of Worms is
   a Eiswein   b Steinwein   c Liebfraumilch   d Landwein
57 The river Rhine begins in
   a Lake Geneva   b Lake Constance   c Lake Balaton   d Lake Maggiore
58 France normally produces, by comparison with Germany
   a 20 times more wine   c a similar amount
   b less wine   d 10 times more wine
59 Slight natural sparkle in German wines is called
   a *schluck*   b *schloss*   c *spritzig*   d *spätlese*
60 Dumpy 'boxbeutel' bottles are used for bottling the wines of
   a Franconia   b Baden   c Nahe   d Württemburg
61 Bingen and Assmanshausen are best known for the production of
   a dry white wine   b red wine   c sparkling wine   d sweet white wine
62 Which of the following is a red wine of Piedmont?
   a Valpolicella   b Barbaresco   c Chianti   d Bardolino
63 Soave is a dry white wine from
   a Latium   b Veneto   c Marches   d Tuscany
64 *Frizzante* wines are produced in Emilia-Romagna known by the name
   a Lambrusco   b Lacryma Christi   c Orvieto   d Asti spumante
65 Italian quality wines are referred to by the use of the initials
   a KWV   b AOC   c DOC   d QmP
66 Verdicchio is a
   a sweet white wine from Puglia   c dry white wine from Marches
   b red wine from Piedmont   d sparkling wine from Tuscany
67 *Classico* means
   a heartland of a region   c wine has been matured in wood
   b best wine of the district   d oldest selection available
68 Marsala wine was originally produced by the English brothers
   a Woods   b Woodhouse   c Wordsworth   d Woodward
69 Lacryma Christi wines are made from grapes grown on the slopes of
   a Stromboli   b Etna   c The Alps   d Vesuvius
70 Romans may go out to the Castelli Romani hillside restaurants to drink the local
   a Frascati   b Soave   c Chianti   d Locorotondo
71 Controls on the wine chianti were first introduced in
   a 1860   b 1910   c 1932   d 1973

72 The 'black cockerel' is the symbol used by the grower's *consortzia* in the classico district of
   a Barolo b Chianti c Valpolicella d Bardolino

73 DOCG wines of Italy are not allowed to be sold in containers which are larger than
   a one litre b five litres c 15 litres d 20 litres

74 The centre of the Italian vermouth trade is the city of
   a Milan b Rome c Naples d Turin

75 More than half of Spain's table wine is produced in the region of
   a La Mancha b Rioja c Penedés d Valencia

76 The main town in La Mancha is
   a Haro b Logrono c Valdepeñas d Córdoba

77 The best quality table wines of Spain are considered to be from Rioja, because of the influence of
   a favourable winds   c French growers
   b southern aspect   d chalk soil

78 Moscatel is a
   a red wine b dry white wine c sweet fortified wine d sparkling wine

79 Spanish table wine with citrus juice is called
   a sangria b tinto c rosado d consecha

80 The Methuen Treaty gave special preference on entry to England to wines from
   a Spain b Portugal c France d Germany

81 Vinho verde means that the
   a wine is green in colour
   b grapes used in the wine were under-ripe
   c wine should be drunk in the year of production
   d maturation has been completed in new casks

82 Mateus rosé is made in the area of
   a Dão b Algarve c Estremadura d Douro

83 Altar wines are produced at
   a Setúbal b Carcavelos c Bucelas d Colares

84 Grapes for Cyprus wines are largely concentrated on the slopes of the mountains known as
   a Troodos b Nicosia c Limassol d Famagusta

85 Commandaria St John is of added interest because it is
   a The most expensive wine from the Mediterranean
   b Matured in wood for 50 years before sale
   c the oldest known named wine in the world
   d all sold to raise funds for a charity

86 Cyprus wines have continued to improve because of the
   a partition of the island   c British departure
   b Turkish influence   d Rossi report

87 The KWV is a
   a private commercial company   c government department
   b wine farmer's association   d trade union

88 Which of the following groups contains only South African wine centres?
   a Pauillac, Cantenac, Pessac   c Toledo, Salamanca, Logrono
   b Paphos, Larnaca, Kyrenia   d Paarl, Worcester, Robertson

89 Van der Hum is a
   a herb liqueur from Australia
   b tangerine flavoured liqueur from South Africa

c New Zealand wine flavoured with herbs
d spirit made from aniseed in California
90 James Busby developed the vineyards of
a Australia b California c Israel d South Africa
91 Rutherglen vines were first planted by
a religous immigrants b convicts c gold prospectors d cowboys
92 The vine root-stocks which are immune to the disease *phylloxera* are grown in the region of
a Maryland b Finger Lakes c Napa Valley d Oregon
93 Which of the following grape varieties is used to make a varietal white wine in California?
a Sultana b Grenache c Cabernet d Zinfandel
94 Premium wines of the USA are loosely considered to be equal to
a AOC b Landwein c Vin de pays d Vin ordinaire
95 The Italian influence on Californian wines is most evident in
a Monterey b Sacramento c Sonoma d Napa Valley
96 The Finger Lakes region of the USA is situated
a in the Napa Valley   c in Washington state
b immediately south of Lake   d 20 miles east of Los Angeles
   Ontario
97 Which of the following is a red wine from Hungary?
a Bardolino b Rioja c Bull's blood d Chianti
98 The fine dry rosé wine from southern Hungary is called
a Pecs b Pradel c Tavel d Balatoni
99 *Monimpex* is
a a material used for wine containers in Hungary
b state control of Hungarian wines
c a Hungarian food which goes well with dry white wine
d the river which runs through Budapest
100 Which of the following lists contains only countries which share a common frontier with Hungary?
a Italy, Turkey, USSR, Albania, Greece
b Germany, Belgium, Denmark, Norway, Sweden
c Israel, Turkey, Greece, USSR, Italy
d Romania, USSR, Yugoslavia, Czechoslovakia, Austria
101 When Tokay is made without the use of 'noble rot' grapes it will be labelled
a *szamorodni* b *puttonyos* c *gonci* d *aszu*
102 A normal bottle of tokay contains
a 70 cl b 75 cl c 50 cl d 60 cl
103 English wine-making on a commercial basis was revived by Guy Salisbury-Jones in
a Devon b Hampshire c Dorset d Kent
104 How many days of sunshine are considered to be necessary between the flowering of the vine and the harvest, to make an acceptable wine?
a 50 b 80 c 100 d 120
105 Moldavia is a wine producing region of
a Switzerland b Germany c USSR d Argentina
106 Mavrodaphne is a rich sweet dessert wine from
a Greece b Israel c Yugoslavia d New Zealand
107 Château Musar is a famous vineyard in
a Israel b Switzerland c Lebanon d Algeria

108 Cook's chenin blanc is a well-known white wine from
    a Australia b New Zealand c United Kingdom d California
109 The South American country which produces most wine is
    a Chile b Paraguay c Brazil d Argentina
110 Glacier wines are those wines of Switzerland which have been
    a refrigerated to remove crystals of tartar
    b made from grapes which were crushed while frozen
    c kept in cold storage at −6 °C for more than one year
    d matured in caves above the snow line
111 The vineyards of Israel owe their development to the wealth of
    a Baron Philip de Rothschild   c Baron Le Roy de Boiseaumarie
    b Count Harasthy   d Lord Cowdray

Part 3   The service of wine

1 When serving wine the bottle neck should be twisted
    a to avoid drips on the table-cloth
    b so that the guest can read the label
    c to send the sediment back down the bottle
    d so that the waiter can see how much is left
2 The bottle of wine is presented to the host so that he can
    a explain the label to his guests
    b check the colour of the wine
    c verify his choice before it is opened
    d read the alcoholic strength
3 When sampling wine, the order of assessment is as follows
    a colour, bouquet, clarity, taste   c bouquet, colour, taste, clarity
    b taste, clarity, colour, bouquet   d clarity, colour, bouquet, taste
4 When removing the cork from sparkling wine the bottle should be held
    a horizontally b vertically c at an angle of 45 ° d upside-down
5 Which of the following groups of glassware could all be used for the service of beer
    a wellington, schooner, thistle, jean
    b tulip, dimple, napoleon, flute
    c nonik, worthington, pilsner, dimple
    d continental, paris, old-fashioned, viking
6 A viking glass would be suitable for
    a sherry b table wine c bottled beer d brandy
7 Brown stemmed glasses should be used for the service of
    a Rheingau and Nahe   c Alsace wines and Moselles
    b Steinweins and Clarets   d Vinho Verdes and Loire wines
8 Liqueur glasses which cannot be washed immediately after use should be
    a filled with water   c left upside-down on a tray
    b placed in a sink full of hot water d left in a cupboard until morning
9 If the customers complain that their red wine is too cold the waiter should
    a place it in the microwave for five minutes
    b pour it into a warmed decanter
    c hold it under hot running water for several minutes
    d leave it upright in the bain-marie for ten minutes
10 The ideal temperature for the service of rosé wines is approximately
    a 10 °F b 65 °F c 10 °C d 65 °C

11 Which of the following alcoholic beverages can be made anywhere in the world without infringement of labelling laws?
a Scotch   b Port   c Champagne   d Brandy
12 Non-vintage wines are wines
a of several years blended together
b from non-quality areas
c of a poor year
d which do not reach 8.5 ° of alcohol by volume
13 *Süss* is the German word for
a medium   b medium dry   c very dry   d sweet
14 German landwein is always
a slightly sparkling   b dry or semi-dry   c sweet   d expensive
15 Italian DOCG wines must be sold by Italian law
a in containers of not more than five litres
b before March in the year of production
c in EEC countries only
d in raffia fiaschi
16 Which of the following wines will need to be decanted?
a Cloroso sherry   b Sercial madeira   c Vintage pòrt   d Moscatel
17 Decide which of these statements is true
a decanters do not have stoppers
b square-sided decanters are intended for spirits
c carafes are usually made of fine crystal
d ship's decanters are fitted with rubber suction pads

Part 4   Fortified wine

1 Shipper's port-wine lodges are to be found in the town of
a Sanlúcar de Barrameda   c Puerto de Santa Maria
b Vila Nova de Gaia   d Jerez de la Frontera
2 A cask of port containing 115 gallons (523 litres) is called a
a pipe   b barrel   c hogshead   d butt
3 Which statement is correct?
a Port is fortified after fermentation is complete
b Port is not a fortified wine
c The fortification of port takes place before fermentation starts
d Port fortification is carried out during fermentation
4 Port is often referred to by the Portuguese as
a 'wine for the American market'   c 'the Englishman's wine'
b 'Frenchman's delight'   d 'Europe's choice'
5 Approximately how much of the wine produced in the Upper Douro is port wine?
a one-tenth   b one-third   c three-quarters   d all of it
6 Tawny port is
a wine of one year, bottled after two years, and ready to drink after 10 years
b wine of one year, but bottled after five years and ready for immediate drinking
c made from white grapes only, and drunk as an aperitif
d a blend of ports left in cask for up to 25 years
7 A *quinta* is a
a vineyard or property in the Upper Douro

   b tool used for pushing the skins down into the fermenting wine
   c foreman in the vineyard
   d boat which was used to take port-wine down the river Douro

8 Amongst the villages of the Upper Douro are
   a Haro, Alfaro, Logrono   c Carcavelos, Setúbal, Colares
   b Yecla, Jumilla, Valencia   d Regua, Villa Real, Pinhao

9 Pedro X is used in the sherry region of Spain to
   a clarify b preserve c sweeten d fortify

10 The layer of yeast which forms on some sherries in the cask is called
   a *bodega* b *albariza* c *flor* d *barros*

11 Which of the following is not a type of sherry?
   a Montilla b Manzanilla c Amontillado d Oloroso

12 A shipping butt of sherry contains
   a 115 gallons b 54 gallons c 108 gallons d 36 gallons

13 The grape variety which is most widely grown to make sherry in Andalucía is
   a Muscat b Palomino c Muscadelle d Garnacha

14 Which of the following rivers flows through the sherry producing region in Spain?
   a Guadalquivir b Minho c Dão d Douro

15 During the second and third years of its life sherry is in the stage known as
   a *criadera* b *añada* c *arenas* d *raya*

16 The soil on the island of Madeira is mainly
   a sandy loam   c fertile potash
   b gravelly pebbles   d absorbent chalk

17 A *pergola* is a
   a support for vines growing high
   b vehicle used to transport grapes
   c skilfully designed irrigation system
   d goatskin bag used to carry grape-juice

18 The capital of Madeira is
   a Malmsey b Zarco c Levada d Funchal

19 Rainwater is a type of
   a Sercial b Verdelho c Bual d Malmsey

20 After treatment in the *estufa*, the wines of Madeira, before sale, must rest for
   a one month b six months c 13 months d four years

21 In 1786 Antonio Carpano produced the first of the modern commercial vermouths. It survives today as
   a Martini b Cinzano c Campari d Punt e mes

22 The centre for the vermouth trade of Italy is
   a Milan b Turin c Naples d Sienna

23 French vermouth is centred on the city of
   a Marseilles b Lyons c Bordeaux d Orléans

24 French vermouth differs significantly from Italian in its manufacture as it is usually
   a fortified with rum   c sweetened with honey
   b matured for five years d weathered for two years

25 In response to a request for a 'gin and French' the bar-tender would be correct to serve gin with
   a Cognac b Perrier c Noilly Prat d Pernod

26 The sweet white fortified wine from California is known as
   a Commandaria b Angelica c VDN d Moscatel

27 Dry marsala is indicated on the label by the word
   a sec   b virgen   c brut   d natur
28 Evaporated unfermented grape-juice which has been cooked over an open fire is used to sweeten
   a Port   b Málaga   c Tarragona   d Moscatel de Setubal

Part 5   Eating and drinking

1 Which of the following alcoholic beverages would be suitable as a preprandial drink at a function?
   a Drambuie   b Barsac   c Sercial   d Monbazillac
2 A suitable wine to accompany roast pheasant would be
   a Muscadet   b Barolo   c Steinwein   d Chablis
3 Entre-deux-mers would be an excellent choice of wine with
   a shellfish   b pasta   c roast lamb   d blue-veined cheese
4 Sweets and desserts would make an excellent combination with
   a Bardolino   b Chinon   c Zinfandel   d Cérons
5 Which of the following wines would you advise the customer to drink with roast leg of lamb?
   a Valpolicella   c Liebfraumilch
   b Soave   d Bernkastler doktor riesling auslese

Part 6   Bar work

1 A six-out measure would be the correct one to use when serving
   a marsala   b gin   c vintage port   d manzanilla
2 Muddlers are used
   a with computer software   c in cellar storage areas
   b to stir cocktails   d when stock-taking in the bar
3 The first president of the United Kingdom bartender's guild worked as a barman at the
   a Ritz   b Savoy   c Metropole   d Dorchester
4 Dry gin, dry vermouth, sweet vermouth and orange juice are the ingredients of a
   a 'Cuba libre'   b 'Highball'   c 'Screwdriver'   d 'Bronx'
5 A 'Harvey wallbanger' cocktail consists of vodka and orange juice topped with
   a Grenadine   b Malibu   c Strega   d Galliano
6 Which of the following cocktails contains fresh cream?
   a 'Royal clover club'   c 'John Collins'
   b 'Brandy Alexander'   d 'Pink lady'
7 Cocktails should be shaken when they
   a contain only clear ingredients   c contain non-alcoholic ingredients
   b have a cloudy ingredient   d include carbonated water
8 Brandy, Cointreau and lemon juice are the ingredients of a
   a 'White lady'   b 'Sidecar'   c 'Screwdriver'   d 'Manhattan'
9 Kir is a mixture of
   a red burgundy wine with cherry brandy
   b white bordeaux wine with cassis
   c dry white burgundy wine with cassis
   d claret with cherry brandy

10 Which of the following cocktails contains alcohol?
   a 'Cinderella' b 'Pussyfoot' c 'Parson's special' d 'Pousse café'

Part 7   Cellar work

1 Racking is the word used by brewers for
   a filling beer casks   c empty crate and bottle procedures
   b placing beer casks on stillions   d correct storage of cellar equipment
2 Beer barrels are prevented from rolling from side to side by the use of
   a shives b scotches c stillions d ropes
3 The beer cellar floor should be made up of
   a quarry tiles b granolithic material c brick d composition cork
4 Cloudy beer will result if the
   a cellar-man has forgotten to remove the spile from the shive
   b cask has been on sale too long
   c cellar contains strong odours
   d beer was delivered in a thunderstorm
5 A firkin of draught beer contains
   a 54 gallons b 36 gallons c 18 gallons d 9 gallons
6 Eighteen gallons of beer in cask is known as a
   a hogshead b kilderkin c barrel d pin
7 Coopers are tradesmen who
   a work in the mill at the brewery
   b make wooden casks from oak
   c turn the barley in the maltings
   d watch over drying hops in the oast-houses
8 Kegs of guinness are different from other kegs of beer in that they
   a are made of shiny metal
   b are not returnable
   c do not need a separate CO_2 cylinder
   d are square in shape

Part 8   Brewing and distilling

1 The finest water for making light ales is found at
   a Chiswick b Burton-upon-Trent c Melton Mowbray d Crewe
2 Barley for brewing is grown in large quantities in
   a Norfolk b Westmoreland c Glamorgan d Kent
3 *Diastase* is the enzyme which
   a breaks down the cell walls in the barley grain
   b causes fermentation to start
   c dissolves the hop flowers in the copper
   d converts the insoluble starch into fermentable sugar
4 Isinglas is used in beer to
   a filter b clarify c flavour d colour
5 The sweet extract of malted barley is boiled in the copper with
   a hops and sugar for about two hours
   b sugar for about six hours
   c hops for one and a half days
   d yeast for about two hours

6 In the brewing process, the stage which is between the mash-tun and the hop-back is the
  a paraflow b copper c fermenting vessel d mill
7 The first sparkling artificial mineral waters were made in 1772 by
  a Betsy Flanagan b Antonio Carpano c Joseph Priestly d Jerry Thomas
8 San pellegrino is a natural mineral water from
  a Italy b France c Belgium d Germany
9 Natural mineral waters from West Germany include
  a Evian b Apollinaris c Vichy Celestins d Perrier
10 The temperature at which alcohol starts to boil is
  a 50 °C b 78 °C c 87 °C d 100 °C
11 The spirit safe is where
  a the distiller keeps his valuable equipment
  b the oldest bottles of liqueur brandy are kept
  c the final blending of spirits takes place
  d heads and tails are separated and testing takes place
12 In the UK the normal rate of evaporation of maturing spirits is approximately
  a 1 per cent p.a. b 3 per cent p.a. c 7 per cent p.a. d 10 per cent p.a.
13 Spirits made by the pot-still method include
  a vodka b brandy c grain whisky d light rums
14 Pot still spirits differ from those produced in a patent still as follows
  a they contain more impurities
  b they contain fewer impurities
  c they do not need to be matured for two years
  d they are much higher in alcohol when distilled
15 The analyser and rectifier are
  a whisky blenders   c employees at the distillery
  b parts of the pot still   d parts of the patent still
16 Gin was first produced in
  a Amsterdam b Geneva c London d Plymouth
17 Oil of juniper was used as a medicine for
  a heart disease   c kidney trouble and gout
  b chest and throat complaints   d headaches and migraines
18 'Dutch Courage' was a name which was given to
  a whisky b brandy c tequila d gin
19 Malt wine is a type of
  a wine b spirit c beer d lager
20 The most common botanical used in the flavouring of gin is
  a coriander b fennel c juniper d bay leaves
21 The region of France where cognac brandy is produced is crossed by the river
  a Charente b Gironde c Dordogne d Seine
22 Grande Champagne is a region of France which is famous for the production of
  a Calvados b Cognac c Champagne d Claret
23 Which of the following is a type of brandy?
  a Arrack b Asbach c Aquavit d Absinthe
24 French law dictates that before sale VSOP brandy must be matured for
  a one year b two years c three years d four years
25 Old VSOP brandy will improve most while it is in
  a cask b bottle c stainless steel d glass fibre
26 Rum is produced from
  a grain b sugar cane products c bamboo d sunflowers

27 Captain Morgan is a rum which is produced in
   a Trinidad   b Jamaica   c Guyana   d Cuba
28 Which of the following is a white rum?
   a Tropicana   b Lemon hart   c Captain Morgan   d Lamb's navy
29 The term *grog* referred to
   a weak rum which was drunk by slaves
   b crushed sugar cane
   c a daily issue of rum and water given to British sailors
   d caramel colouring used in dark rums
30 Grain whisky is distilled in Scotland in a patent still from
   a barley   b rye   c wheat   d maize
31 De luxe whiskies are those which are
   a matured longer than usual   c sweetened with honey
   b higher in alcohol   d sweetened with cane sugar
32 Speyside whisky distillers attribute the quality of their product to the fact that
   the water in the burns tumbles over
   a quartz pebbles   b yellow gravel   c white marble   d red granite
33 Highland malt whiskies are separated from the Lowland malts by an imaginary
   line which runs from
   a Aberdeen to Perth   c Dundee to Greenock
   b Perth to Oban   d Edinburgh to Ayr
34 Campbeltown malt whisky distilleries are on the
   a Mull of Kintyre   c Isle of Islay
   b Kyle of Lockalsh   d shores of Loch Ness
35 Scotch whisky owes its predominance over brandy on world markets to
   a trade treaties   c scotsmen living abroad
   b low prices   d *phylloxera* in the vineyards
36 Irish whiskey differs from Scotch during manufacture as follows
   a barley in the maltings does not come into contact with peat smoke
   b it is distilled at a higher temperature than Scotch
   c the grain is roasted before it is fermented
   d special flavourings are added after distillation
37 Old crow is made by pot-still methods in
   a Ireland   b Scotland   c Bourbon county   d Canada
38 Which of the following are *all* American Bourbon whiskies?
   a John Power, Jim Beam, highland park, four roses
   b Old cameron brig, Jamesons, old Kentucky, wild turkey
   c Jack Daniels, Jim Beam, old grandad, four roses
   d Old hickory, Jack Daniels, yellowstone, old bush mills
39 Canadian club is the best-known Canadian whiskey. The predominating
   cereal is
   a maize   b rye   c wheat   d barley
40 The name Calvados came originally from
   a the name of the man who first made a spirit from apples
   b an apple variety
   c one of the Normandy villages
   d a Spanish galleon wrecked off the coast
41 The finest Calvados spirit is known as
   a Calvados reglémentée   c eau de vie de cidre
   b pays d'auge   d un trou Normand

42 Mirabelle is a colourless spirit made from
   a pears b cherries c plums d peaches
43 Which of the following pairs of spirits both turn milky with water?
   a Quetsch and Sake   c Tequila and Slivovitz
   b Ouzo and Pernod   d Kirsch and Vodka
44 The spirit which is traditionally served in warm porcelain cups is
   a Schnapps b Sake c Pernod d Raki
45 Salt and a wedge of either lime or lemon should be provided with
   a Tequila b Poire William c Framboise d Arrack
46 Tequila is distilled from
   a nuts b fennel twigs c cactus plants d bananas
47 Vodka is purified by being passed through
   a chalk-dust b activated charcoal c a centrifuge d a paraflow
48 Calypso coffee contains
   a Tia Maria b crème de cacao c dark rum d light rum
49 Cognac brandy should be used in
   a highland coffee b witch's coffee c monk's coffee d café royale
50 Caribbean coffee is made with the inclusion of
   a Tequila b Rum c Tia Maria d Bourbon
51 The main flavouring ingredient of Kümmel is
   a juniper b caraway c aniseed d coriander
52 Which of the following glasses would be most suitable for the service of a liqueur?
   a Wellington b Elgin c Balloon d Pilsner
53 Italian liqueurs include
   a Strega b Bénédictine c Curaçao d Tia Maria
54 The liqueur which has the highest alcoholic strength is
   a Grand Marnier b Galliano c Yellow Chartreuse d Green Chartreuse
55 One of the most popular herb flavoured liqueurs is
   a Bénédictine b Cointreau c Kahlúa d Amaretto di Saronno
56 Southern comfort is made in the state of
   a Missouri b Alabama c Kentucky d Virginia
57 Crème de framboises is flavoured with
   a strawberries b blackcurrants c gooseberries d raspberries
58 Which of the following liqueurs is bright yellow in colour?
   a Kümmel b Strega c Brontë d Cointreau
59 Which of these liqueurs has an Irish connection?
   a Carolans b Merlyn c Athol brose d Brontë
60 Midori is flavoured with
   a kiwi fruit b melon c herbs d coconut milk
61 A liqueur which is produced in the Cognac area of France is
   a Parfait Amour b Van der Hum c Maraschino d Grand Marnier
62 Forbidden fruit liqueur is noted for its
   a high alcoholic strength   c attractive bottle
   b keeping qualities   d production in a range of colours
63 Sapin d'Or is produced in a bottle which is
   a frosted b square-shaped c like a church window d log-shaped
64 The colour of Parfait Amour may be
   a green or colourless   c purple or pink
   b yellow or brown   d orange or blue

65 Advocaat is made from
 a white wine, egg yolks, sugar   c honey, white wine, whole egg
 b whole egg, brandy, honey   d brandy, egg yolks, sugar
66 Kahlūa is flavoured with
 a cherries including the stones   c peach and apricot kernels
 b Mexican coffee beans   d mandarin oranges
67 If the cocktail bar-tender required a colourless liqueur with an orange flavour, he could use
 a Edelweiss   b Cointreau   c Crème de Noyau   d Maraschino
68 Potential alcoholic strength can be measured by placing the juice of a grape into a
 a barometer   b refractometer   c chronometer   d hydrometer
69 A bottle of whisky at drinking strength contains
 a 40 per cent water   c 70 per cent alcohol
 b 60 per cent water   d 30 per cent alcohol
70 The Sikes scale of alcoholic strength was developed from the
 a different weights of alcohol and water
 b ignition of gunpowder with spirit
 c sugar content of alcohol
 d boiling points of alcohol and water
71 Pure alcohol on the Sikes scale is approximately
 a 100 °   b 157 °   c 175 °   d 200 °

Part 9   Storage of alcohol and other items

1 Red wines will keep longer if there is a high content of
 a tartaric acid   b tannin   c sugar   d sediment
2 Smoking in England was made fashionable by
 a Christopher Columbus   c John Bunyan
 b Sir Walter Raleigh   d John Woodhouse
3 The first cigarettes, called papelettes, were made in
 a Spain   b France   c Holland   d Italy
4 Briar pipes are made of wood cut from
 a brambles   b willow   c roses   d heather
5 In Britain, the city which is most involved in the tobacco trade is
 a Liverpool   b Glasgow   c Bristol   d Manchester
6 Black tobacco, called *Balkan Sobranie* is grown in
 a Florida, Pennsylvania, Wisconsin   c Kentucky, Louisiana, Tennessee
 b Georgia, Ohio, Florida   d Romania, Greece, Turkey
7 The best quality cigar tobacco is grown in
 a Cuba   b Zimbabwe   c Yugoslavia   d Syria
8 The names panatella and cheroot are used to indicate
 a trade names   b sizes of cigars   c district of origin   d tar content
9 Upon import, tobacco must remain in a bonded warehouse for
 a one year   b two years   c three years   d four years
10 When serving a corona cigar the waiter should
 a cut the cigar band neatly   c leave the cigar band in position
 b slide the band off the cigar   d open the band where it is glued

Part 10   Customer relations

1 When a customer enters your establishment you should
  a   try to sell them your most expensive products all evening
  b   greet and welcome them; sell your products with tact
  c   ignore them
  d   refuse to serve difficult customers
2 Wine lists should be
  a   as large and ornate as possible
  b   *never* taken away by the customer
  c   clean and presentable; clearly and logically set out
  d   listed so that aperitifs fall on the last page

Part 11   The law, health and safety

1 A notice must be prominently displayed in the bar stating which measure is being used for the service of
  a   Brandy, Rum, Gin, Vodka   a   Whisky, Gin, Rum, Vodka
  b   Vodka, Brandy, Rum, Whisky   b   Gin, Vodka, Whisky, Brandy
2 Under the Price Marking (Food and Drink on Premises) Order 1979 prices must be prominently displayed to be seen by a customer before reaching the area of consumption for
  a   four wines   b   six wines   c   eight wines   d   10 wines
3 The legal limit beyond which it is against the law to be in charge of a vehicle is
  a   25 microgrammes of alcohol in 100 ml of breath
  b   35 microgrammes of alcohol in 100 ml of breath
  c   25 milligrammes of alcohol in 100 ml of blood
  d   35 milligrammes of alcohol in 100 ml of blood
4 A child under 14 may be allowed in a licenced bar if
  a   he is with his parents   c   it is a bank holiday
  b   he is drinking only mineral waters   d   his parents are residents
5 If a drunken person is asking for a double gin but is not causing trouble or being a nuisance, you should
  a   allow him just one more drink and warn him that it will be the last
  b   serve him and ask him to sit quietly in the corner
  c   refuse to serve him
  d   eject him by force into the car-park
6 Residents on licenced premises
  a   must be served with alcohol at any time of the day or night if they demand it
  b   can entertain just one single guest after normal hours so long as the resident pays for the drinks
  c   may be served with alcohol after hours if the management are willing to serve them
  d   can entertain an unlimited number of guests so long as the guests pay for all the drinks
7 Exercises in fire drill should by law be carried out once in every
  a   quarter   b   half-year   c   year   d   two years
8 Water jet extinguishers should be used when fighting fire in
  a   electrical appliances   b   gas appliances   c   fat fryers   d   clothing stores
9 Halon extinguishers are coloured
  a   green   b   black   c   blue   d   white

10 'Nippon' is a proprietary treatment for infestation by
    a wood-lice b ants c earwigs d moths
11 Which of the following pests are nocturnal?
    a Silverfish b Spiders c Cockroaches d Black ants
12 Spiders like a habitat which is

a dark and damp	c humid with decaying food
b wet and rotting	d dry and warm

13 Silverfish thrive on a diet of

a flour and paper	c rotting wood
b decaying meat products	d fruit and vegetables

14 Bar swabs should be left soaking overnight in

a warm soapy water	c weak bleach solution
b mild detergent	d mild disinfectant

(Multiple-choice) Points possible 280   Points obtained ____

Total points possible 1600   Points obtained ____

Section twelve
Answers

Section one   Wine from many countries

1 French wine and spirit regions
1 Calvados 2 Champagne 3 Alsace 4 Loire 5 Chablis 6 North Burgundy
7 South Burgundy 8 Jura 9 Cognac 10 Savoie 11 Bordeaux 12 North Rhône
13 South Rhône 14 Armagnac 15 Midi 16 Provence

2 Wine regions of Burgundy
1 Chablis 2 Dijon 3 Gevrey-Chambertin 4 Côte d'Or 5 Beaune 6 Pommard
7 Volnay 8 Meursault 9 Puligny-Montrachet 10 Santenay 11 Mercurey
12 Montagny 13 Tournus 14 Cluny 15 Pouilly-Fuissé 16 Mâcon 17 Juliénas
18 Bourg 19 Chénas 20 Moulin-à-Vent 21 Fleurie 22 Morgon 23 Brouilly
24 Lyon 25 Beaujolais

3 French wine
1 Claret 2 Sauternes 3 Entre-deux-Mers 4 Eleanor of Aquitaine 5 Médoc
6 Château Latour, Château Margaux, Château Lafite-Rothschild or Château Haut-Brion
7 French Revolution 8 Hospices de Beaune. Auction. November 9 Beaujolais
10 Pinot Noir. Gamay 11 Burgundy 12 White. Dry 13 Tavel
14 Lavender. Thyme 15 Red 16 Vosges 17 Ill 18 Edelzwicker 19 Vouvray
20 Dry 21 Provence 22 Sparkling 23 Jura 24 Touraine 25 Bottled
26 Hermitage 27 Gamay 28 Pinot Noir. Chardonnay 29 Cabernet Sauvignon
30 Mistral 31 Châteauneuf-du-Pape 32 Côte de Nuits. Côte de Beaune
33 Castillon 34 Anjou-Saumur 35 Jura 36 Rosé (pink)

4 German wine regions
1 Mittel-Rhein 2 Ahr 3 Mosel-Saar-Ruwer 4 Rheingau 5 Nahe 6 Rheinhessen
7 Hessische Bergstrasse 8 Franconia 9 Rheinpfalz 10 Baden 11 Württemburg

5 German wine and spirits
Vertical words   SPRITZIG, LIEBFRAUMILCH, QUALITATSWEIN, SEKT, KABINETT,
TROCKEN, FRANCONIA
Horizontal words   EDELFAULE, SCHLOSS, PALATINATE, WORMS, HAUPTLESE,
NAHE, MOSEL, TAFELWEIN, BADEN
Diagonal words   RIESLING, TROCKENBEERENAUSLESE, AUSLESE, ROTWEIN,
AHR, EISWEIN, SPATLESE, ASBACH, BINGEN

6 Italian wine regions
1 Turin 2 Vermouth 3 Piedmont 4 Asti Spumante 5 Barolo 6 Soave
7 Veneto 8 Valpolicella 9 Bardolino 10 Lambrusco 11 Emilia Romagna
12 Verdicchio 13 Marches 14 Tuscany 15 Chianti 16 Umbria 17 Orvieto
18 Rome 19 Frascati 20 Locorotondo 21 Lacryma Christi 22 Naples
23 Campania 24 Marsala 25 Etna Rosso 26 Etna Bianco

7 Italian wine
Vertical   AMABILE, BARDOLINO, EST, EST, EST, BARBERA, BARBARESCO,
VALPOLICELLA, VERDICCHIO, ASTI SPUMANTE, ETNA ROSSO

Horizontal   FRIZZANTE, LOCOROTONDO, NEBBIOLO, GOVERNO, SOAVE,
LACRYMA CHRISTI, FRASCATI, MARSALA
Diagonal   ORVIETO, CHIANTI, FIASCHI, LAMBRUSCO, BAROLO, BIANCO,
CLASSICO, SECCO

8 *Champagne and sparkling wine*
 1 Magnum   2 Chalk   3 (a) Pinot Noir (b) Pinot Meunier (c) Chardonnay
 4 Dom Pérignon   5 Black   6 Hautvillers   7 Bouzy   8 Swizzle-stick
 9 Marne. Aube   10 Blanc des Blancs   11 26   12 4000 kilos   13 Vin de cuvée
14 *Rebêche*   15 *Assemblage*   16 *Rémuage*   17 *Dégorgement*   18 *Dosage*
19 Brut. Natur   20 Rich. Doux   21 Flute   22 Four   23 Tank   24 45 °
25 7 or 8 °C   26 Impregnation   27 Saumur. Vouvray   28 Blanquette de Limoux
29 Germany   30 Asti spumate

9 *Spanish wine*
Vertical   PALOMINO, FLOR, RIOJA, JEREZ DE LA FRONTERA, EBRO,
CONSECHA, MOSCATEL, ALELLA
Horizontal   LA MANCHA, CATALONIA, BUTT, BODEGA, TINTO, TINAJARS,
PENEDES, MANZANILLA, MALAGA, ESPUMOSO, TARRAGONA
Diagonal   DULCE, SAN SADURNI DE NOYA, MONTILLA, SANGRIA, SOLERA,
BLANCO, ESPARTO

10 *Wine, other than French*
 1 Harvest   2 30 ° and 50 °   3 Tannin   4 Copper sulphate   5 Hambledon
 6 Caves   7 California   8 *Phylloxera*   9 Grafting   10 Australia   11 KWV
12 Huguenots   13 Refrigeration   14 Vinho verde   15 Spain   16 Red
17 La Mancha   18 Sangria   19 Lacryma Christi   20 Verdicchio or soave
21 Governo   22 Black cockerel   23 Frascati   24 Beerenauslese   25 Steinwein
26 Worms   27 Slight sparkle   28 *Rotwein*   29 Müller–Thurgau   30 Ahr
31 Palatinate

11 *Wine (from many countries)*
 1 True   2 False   3 False   4 True   5 False   6 False   7 False   8 True   9 False
10 False   11 False   12 False   13 True   14 False   15 True   16 True   17 False
18 False   19 False   20 True   21 False   22 True   23 True   24 True   25 False

12 *Colour of wine*
 1 white   2 red   3 red   4 rosé   5 both   6 rosé   7 red   8 white   9 white
10 red   11 red   12 white   13 red   14 red   15 white   16 white   17 rosé
18 white   19 white   20 red   21 white   22 red   23 red   24 red   25 white
26 red   27 red   28 white   29 red   30 both   31 rosé   32 red   33 red   34 white
35 white   36 rosé   37 red   38 white   39 white   40 red

13 *Annual production*
Italy 1500 million gallons, France 1400, USSR 700, California 550, Spain 440, Portugal
220 and Germany 70.

Section two   The service of wine

1 *Case studies on table service*
 1 Joanne told Stephen who quietly explained that the policy of the Far Forest Restaurant
   was to charge a small amount of 'corkage' on each bottle of wine, but that he would be
   pleased to chill and serve the wine for them.
 2 Stephen correctly went ahead and served the red wine with the fish course without com-
   ment. The customer is paying for the wine and must be allowed to choose what you and
   many other people would consider to be the wrong wine with a particular food.
 3 Stephen had been well trained. He quietly took the bottle of wine, and poured it into a
   decanter which he had warmed gently under the hot tap. This is an acceptable method as

the wine slowly reaches the *chambré* temperature. Later he told Mr Steele that he would never put a bottle of wine in the microwave, or use any other artificial means to alter its character and nature.

4 Stephen spoke quietly to the host asking if one bottle would be sufficient. When the host said that it would, he went ahead and portioned the wine equally into all the glasses.

5 Using his *tastevin* Stephen tasted the wine himself, removed it from the table and served another bottle. The original bottle would be returned to the supplier for credit as *ullage*. A good sommelier would replace the bottle even if he himself could detect nothing wrong with the wine, thereby appealing to the better nature of the customer who, because of this event, would be considered by his guests to have good knowledge of wine.

6 On this occasion, although Stephen fully understood that it is better to open champagne quietly, he allowed the cork to 'pop' (making sure of course that he was in full control of it) in the palm of his hand to avoid injury to himself or other diners.

7 Joanne immediately sent a message to Mr Steele. She asked the man for proof of identity and wrote down his particulars. Mr Steele was not satisfied that the man's behaviour was genuine so he called the police. Usual practice is to ask the person for the telephone number of someone who could be contacted to substantiate the person's identity.

8 As this could have been a genuine absent-minded mistake, Samantha took particulars, in case the coat was returned by another customer. She also obtained a description of the coat, and telephoned as many of the other diners as possible later that day. She informed the guest that the police would be notified, but that the restaurant would reimburse him fully if the coat was not found.

9 The old red wine was carefully decanted in full view of the guests after presentation in its dirty state, again into a warmed decanter. The dirty bottle and the cork were then placed on the table, in a cradle, along with the decanted wine; Stephen then served the wine with the appropriate course.

10 When the Professor left the table Stephen retrieved and re-corked the bottle of burgundy. He wrote the guest's name and room number on the label, and made sure that the wine was taken up to the Professor's room. He informed the manager that the guest was ill in case a doctor should be required.

2 Glassware recognition
1 (from left to right): (c) (h) (e) (f) (a) (i) (d) (g) (b)
2 (from left to right): (c) (d) (g) (b) (f) (h) (e) (a)
3 (from left to right): (d) (f) (b) (e) (a) (c)
4 (from left to right): (c) (a) (d) *(or (d) (a))* (e) (b)
5 (from left to right): (b) (c) (a)
6 (from left to right): (c) (a) (b)

Section three   Fortified wine

1 Fortified wine from many countries A
1 Brandy 2 Tawny 3 Tarragona 4 Woodhouse. Sicily 5 Sercial 6 *Solera*
7 Vintage. Crusted 8 Butt 9 *Quintas* 10 Palomino 11 Cognac 12 Mission
13 Dry. White 14 Antonio Carpano 15 Estufa 16 115 gallons 17 Cyprus
18 Wormwood 19 Madeira 20 Flor 21 Douro. March. Vila Nova de Gaia
22 Potash. Fire 23 Virgen 24 Any one of the following: Rivesaltes, Banyuls,
Frontignan, Rasteau, Beaumes-de-Venise 25 France 26 Bodega 27 Left 28 Finos
29 20 30 Three-out 31 Manzanilla 32 *Oloroso* 33 *Levadas*
34 Pedro Ximénez (Pedro X) 35 Portugal 36 *Anada* 37 Marseilles 38 Esparto
39 During 40 *Albariza, barros* or *arenas* 41 *Criadera* 42 10 43 Amontillado
44 Venencia

2 Fortified wine from many countries B
1 False 2 False 3 True 4 True 5 False 6 True 7 False 8 True 9 True
10 False 11 False 12 True 13 False 14 True 15 True 16 False 17 True
18 True 19 False 20 False 21 True 22 True 23 False 24 True 25 False

Section four   Eating and drinking

1 *Wine with food A*
1 false   2 true   3 false   4 true   5 false   6 false   7 false   8 true   9 true
10 true   11 true   12 false   13 false   14 false   15 true   16 false   17 false
18 true

2 *Wine with food B*
1 Châteauneuf-du-Pape   2 Chablis   3 Vouvray Valentine   4 Chinon   5 Pomerol
6 Médoc   7 Fine old tawny port   8 Tokay Aszu   9 Liebfraumilch
10 Bernkastler Riesling   11 Malmsey Madeira   12 Tio pepe fino sherry   13 Barsac
14 Rioja tinto   15 Entre-deux-Mers   16 Chianti classico   17 Anjou rosé
18 Malvern water   19 Riesling d'Alsace   20 Bull's blood   21 Gevrey-Chambertin
22 Perrier water   23 Piesporter Michelsberg   24 St Emilion   25 Sercial Madeira
26 Beaujolais   27 Cérons   28 Sauternes   29 Copper beech sherry   30 Muscadet

Section five   Bar work

1 *Case studies on bar work*
1 As it is illegal to serve alcohol to a person who is already drunk, Charles explained politely and offered the man coffee or a non-alcoholic drink.
2 Charles immediately telephoned the manager in case it was an explosive device, and carried the parcel to a safe place away from the public access. The police were then called but service continued. Charles was in fact criticised by the police for moving the parcel.
3 Charles did not change the cask but tactfully offered bottled alternatives. By so doing he was able to serve everyone before closing time.
4 Charles had been taught to add up the drinks in his head as he served them. When interrupted, he made a note of the total to that point on a handy pad. He then served other people and continued with the original order when the man returned.
5 Charles had left the five-pound note on the till and politely explained that she had made a mistake. As Mrs Walters did not believe him, Charles asked her if she would contact the manager.
6 As the beer must have been purchased by someone else, Charles questioned the boy and explained that people under 18 could not consume alcohol in the bar. He told them that if such a thing happened again they would not be welcome in the White Hart.
7 Charles did not say that they were in the bar. He asked the caller for his name and said that he would go and find out if they were in. He then approached Mr and Mrs Roberts to ask what they wanted him to say.
8 Charles placed the camera in a safe place and wrote a note explaining the time and where it was found, with names of any people he had seen in that area. Later he wrote a note to display, saying 'Camera found – Contact manager'.
9 Charles quickly opened another bottle and placed the broken one aside to be returned to the supplier as *ullage* for credit.
10 Charles explained that it was in order for friends of residents to drink with them after the bar was closed, but that the resident himself must make the purchase. He served the crisps as that is not against the law.

2 *Bar work and taking orders*
1 false   2 true   3 false   4 true   5 false   6 false   7 false   8 false   9 true
10 false   11 false   12 true   13 false   14 false   15 true   16 true   17 true   18 true

3 *Cocktail bases*
1 (a)   2 (c)   3 (e)   4 (b)   5 (a)   6 (b)   7 (h)   8 (d)   9 (e)   10 (h)   11 (b)
12 (e)   13 (a)   14 (h)   15 (g)   16 (d)   17 (b)   18 (f)   19 (b)   20 (f)

Section six   Cellar work

1 Equipment, routines and beer containers
1 true **2** false **3** true **4** true **5** true **6** false **7** true **8** false **9** false **10** true
11 false **12** true **13** true **14** false **15** true **16** false **17** false **18** false

Section seven   Brewing and distilling

1 Brewing commodities and process
1 Burton-on-Trent **2** Barley **3** Norfolk, Suffolk, Yorkshire or Hampshire **4** Invert
5 Primings **6** Hemp (or nettle) **7** Tannin (or lupalin) **8** Pockets **9** Isinglas
10 Oast **11** Kent (Or Hereford and Worcester) **12** Cell division **13** Lager **14** Mill
15 Wort **16** $1\frac{1}{2}$ or two hours **17** three or four. Days **18** Cattle food **19** fertilizer
20 Paraflow. 30 °C

2 Brewing process
1 From top to bottom 1 mill 2 mash-tun 3 copper 4 hop-back 5 wort receiver
6 paraflow 7 fermenting vessel 8 storage tanks **2** (a) cattle cake (b) spent hop fertilizer
(c) surplus yeast **3** (a) liquor (water) (b) (malted) barley (c) sugar (d) hops (e) yeast
(f) isinglas

3 Spirits
1 78 °C **2** Calvados **3** Apples (or pears). Normandy **4** No **5** Brandy. Rum
6 Cuba **7** 40 per cent **8** Tequila **9** Plums **10** Activated charcoal **11** Sake
12 Poteen **13** Pernod. Ouzo **14** Palm trees **15** Wormwood
16 Hangovers. Stomach upsets **17** Black Forest **18** *Phylloxera* **19** Rum
20 Verschnitt **21** Cognac **22** Gin **23** Bourbon **24** Gin **25** Holland **26** Three
27 Juniper (or coriander) **28** Maize **29** Rum **30** Coffey **31** Dundee. Greenock
32 Spey **33** Pays d'Auge **34** Rum **35** Brouillis **36** Rum **37** Caramel
38 Tropicana, white diamond, dry cane, bacardi or daquiri **39** Rye **40** Patent
41 1860 **42** Campbeltown **43** Asbach **44** Limousin **45** Heads. Tails **46** Italy
47 Potatoes

4 Raw materials of spirits
1 Cactus **2** Plums **3** Palm sap **4** Apples (or pears) **5** Potatoes **6** Cherries
7 Grain **8** Barley **9** Wine **10** Pears **11** Wine **12** Rice **13** Sugar-cane **14** Rye
15 Potatoes **16** Raspberries **17** Plums **18** Potatoes (or grain) **19** Sugar-cane
20 Wine **21** Maize **22** Wine **23** Mixed cereals **24** Barley **25** Sugar-cane
26 Plums

5 Beer, cider and spirits
1 True **2** False **3** False **4** True **5** True **6** False **7** False **8** True **9** False
10 False **11** True **12** True **13** False **14** False **15** False **16** False **17** True
18 True **19** False **20** True **21** True **22** True **23** False **24** True **25** True

6 Speciality coffees
1 Irish whiskey **2** Rum **3** Tia Maria **4** Vodka **5** Scotch malt whisky
6 Bénédictine **7** Cognac brandy **8** Cognac brandy **9** Strega
10 Brontë liqueur **11** Tequila **12** Southern Comfort **13** Drambuie

7 Classification of liqueurs
1 (c) **2** (a) **3** (b) **4** (e) **5** (g) **6** (a) **7** (a) **8** (b) **9** (b) **10** (a) **11** (f)
12 (e) **13** (c) **14** (h) **15** (b) **16** (a) **17** (a) **18** (b) **19** (b) **20** (c) **21** (e)
22 (f) **23** (a) **24** (a) **25** (a) **26** (b) **27** (a) **28** (a) **29** (h) **30** (d) **31** (b)
32 (b) **33** (a) **34** (a) **35** (a) **36** (a)

8 Cocktails and liqueurs
1 Cloudy 2 Stirred 3 Sweet vermouth 4 Gin 5 Lime juice 6 Galliano
7 Pineapple juice 8 Cointreau 9 Angostura bitters 10 Tequila sunrise
11 Brandy 12 Vodka. Orange juice 13 Gin. Brandy 14 Cassis 15 Horse's neck
16 Muddler 17 Alexander 18 Green chartreuse 19 Caraway seeds 20 Brontë
21 Strega 22 Colourless 23 Fraisia or crème de fraises 24 Advocaat 25 Melons
26 Violet or pink 27 Rum. Coffee beans 28 Drambuie
29 Edelweiss (or Fior d'Alpi) 30 Tia Maria

9 Temperatures and strengths
1 (d) 2 (a) 3 (i) 4 (h) 5 (j) 6 (b) 7 (c) 8 (e) 9 (g) 10 (f)

Section eight   Storage of alcohol and other items

1 Storage, control and tobacco
1 Brown 2 Upright 3 Direct sunlight 4 Floor 5 White paint 6 20 7 40. 100
8 Two. Seven 9 45 °F (8 °C). 50 °F (10 °C) 10 65 °F (17 °C) 11 Annual
12 'Z' 13 Columbus 14 Erica (or heather) 15 Spain 16 Cuba
17 Balkan Sobranie 18 Cedar 19 Two 20 Opening

Section nine   The law, health and safety

1 Law, fire, pests, cleaning, health and safety
1 False 2 True 3 False 4 False 5 True 6 True 7 True 8 False 9 False
10 True 11 False 12 True 13 False 14 True 15 False 16 False 17 True
18 True 19 False 20 True

Section ten   General section

1 Grape varieties
1 (v), (xi), (g) 2 (f), (l) 3 (vii), (ix), (xii) 4 (a), (b) 5 From: (c), (e), (o), (m), (v)
6 From: (iv), (v), (g), (j) 7 (i), (x), (c) 8 (iii), (c), (m) 9 (ii), (viii), (n)
10 From: (vi), (vii), (viii), (b), (g) 11 From: (iv), (vii), (xii), (b), (h) 12 (d) 13 (k)
14 (iv)

2 Capacities and measures
1 108 gallons 2 115 gallons 3 16 servings 4 32 servings (using six-out measure)
5 50 cl 6 180 gallons 7 2 bottles 8 18 gallons 9 70 cl 10 90 gallons
11 4 bottles 12 36 gallons 13 100 cl 14 9 gallons 15 20 bottles 16 54 gallons

3 Geographical locations
1 Côte des Blancs 2 Andalucía 3 Upper Douro 4 Grande Champagne
5 Rhône valley 6 Côte de Beaune 7 River Gironde 8 Vosges mountains
9 Normandy 10 Vaud 11 Hungary 12 Saar and Ruwer 13 Rheingau
14 River Ebro 15 Tuscany 16 Veneto 17 Campania 18 Hampshire
19 California 20 South Africa 21 Australia 22 Funchal 23 Troodos Mountains
24 Amsterdam 25 Speyside 26 Turin 27 Bahamas 28 Trinidad 29 Yugoslavia
30 River Minho 31 Latium 32 Franconia 33 River Saône 34 Savoie 35 Jura
36 Provence 37 River Rhine 38 Touraine 39 Fleurie 40 Zimbabwe 41 River Ill
42 Midi 43 Yorkshire

4 Alphabet quiz
1 Albariza 2 *Botrytis* 3 Caraway 4 *Dulce* 5 Erica 6 Flute 7 Geneva
8 Hogshead 9 Islay 10 Juniper 11 Kabinett 12 *Levadas* 13 Methuen
14 Nahe 15 Ouzo 16 *Phylloxera* 17 *Quinta* 18 *Rémuage* 19 Szamorodni
20 Tartaric 21 Underberg 22 Vouvray 23 Wormwood 24 Ximinéth 25 Yeast
26 Zinfandel

5 Initials quiz
1 (a) Very Special Old Pale (b) Old matured Cognac or Armagnac   2 (a) Ko-operatieve
Wijnbouwers Vereniging (b) South African wine-farmer's association
3 (a) Appellation Origine Contrôlée (b) Geographical guarantee of French wines
4 (a) Qualitätswein mit Prädicat (b) Distinguished quality German wine
5 (a) Organisation Internationale Météorologique Légale (b) European standard
measurement of alcoholic strength   6 (a) Vin Délimité de Qualité Supérieure
(b) Second quality wines of France from minor districts   7 (a) Denominazione di
Origine Controllata (b) Italian quality control for typical wines of regions   8 (a) Carbon
dioxide gas (or carbonic acid gas) (b) The gas which is given off during fermentation
9 (a) Late Bottled Vintage (b) Port wine of one year bottled after about five years and
ready for drinking   10 (a) Vins doux naturels (b) French fortified wine
11 (a) United Kingdom Bartender's Guild (b) Open to suitably trained bar-tenders who
work regularly in bars   12 (a) Campaign for Real Ale (b) To encourage the brewing,
conditioning and serving of beer in the traditional way   13 (a) Non-vintage
(b) Wines of several years blended together   14 (a) Licensed Victuallers Association
(b) Works for the benefit of the local licencees   15 (a) Hotel, Catering and Institutional
Management Association (b) Professional association of caterers and accommodation
managers   16 (a) Château bottled (b) French wine bottled by the grower
17 (a) Qualitätswein bestimmte Anbaugebiete (b) Quality German wines produced in
specific regions

6 Initials and clues
 1 Licensed Victuallers Association   2 Hotel, Catering and Institutional Management
Association   3 Free house   4 *Vitis vinifera*   5 Copper sulphate   6 Foulloir-egrappoire
 7 Dom Pérignon   8 Montagne de Rheims   9 Méthode champenoise
10 Eleanor of Aquitaine   11 Cabernet Sauvignon   12 Battle of Castillon
13 French Revolution   14 Beaujolais nouveau   15 Hospices de Beaune
16 Lavender and thyme   17 Pouilly-sur-Loire   18 Louis Pasteur   19 Vin jaune
20 Qualitätswein mit prädikat   21 Est, Est, Est   22 Jan Van Riebeeck
23 Little Karroo   24 James Busby   25 Gold rush   26 English Vineyards Association
27 Three Choirs   28 Diethylene glycol   29 Glacier wines   30 Mise-en-place
31 Vila Nova de Gaia   32 Upper Douro   33 *Solera* system   34 *Estufa* system
35 Pineau des charentes   36 Entre-deux-Mers   37 Highly flavoured meats
38 Worthington white shield   39 Betsy Flanagan   40 Al Capone   41 Muddler spoon
42 Dip-stick   43 Chloride of lime   44 Oast house   45 *Saccharomyces carlsbergensis*
46 Wort receiver   47 San pellegrino   48 Joseph Priestly   49 Spirit safe
50 Aeneas Coffey   51 Customs and Excise officer   52 Highly rectified
53 Sylvius van Leyden   54 Cold compounded gin   55 Grande Champagne
56 Limousin oak   57 Rum verschnitt   58 White diamond   59 United Rum Merchants
60 Admiral Vernon   61 Jack Daniels   62 Old bush mills   63 *Un trou Normand*
64 Activated charcoal   65 Café royale   66 Green chartreuse   67 Vieille cure
68 Forbidden fruit   69 Southern comfort   70 Bailey's Irish cream
71 Potential alcohol   72 Under proof   73 Chill haze   74 Credit note   75 Bin-card
76 Christopher Columbus   77 Great plague   78 Bonded warehouse
79 Cedar wood boxes   80 Personal presentation   81 Team-work
82 Sex Discrimination Act   83 Blood alcohol level   84 Fire blanket
85 Dry powder extinguishers   86 Black ants   87 Curriculum vitae

7 Companies
 1 Champagne   2 Ginger wine   3 Gin   4 Rum   5 Bourbon whiskey
 6 Cherry brandy   7 Cognac   8 Mineral waters   9 Cider   10 Beer   11 Sherry
12 Real ale   13 Fruit juices   14 Advocaat   15 Scotch whisky   16 Sherry   17 Cigars
18 Lager   19 Port   20 Beer   21 Irish whiskey   22 Lime cordial   23 Cognac
24 Vodka   25 Gin   26 Champagne

8 Speed test
 1 Californian fortified wine   2 Warehouse for sherry storage
 3 Dessert wine from Cyprus   4 Topping up champagne bottles
 5 Noble rot in German   6 Cognac vine growing district   7 Ground malted barley
 8 Abbey in Champagne district   9 Clarification liquid for beer
10 Equals four champagne bottles   11 White burgundy with blackcurrant
12 Sweet white bordeaux wine   13 Sherry from near sea   14 Very dry champagne
15 Bourbon from pot still   16 Alcohol ban in USA   17 Spirit from plums
18 Medium dry Madeira wine   19 Stirs bubbles from champagne
20 Earthenware wine storage jars   21 Region of Italy   22 Aromatised fortified wine
23 Inventor of marsala   24 Hennessy old matured cognac   25 Blended wine from
Alsace

9 Crosswords

Crossword 1
Across 1 April 5 Metal 8 Nosed 9 Lined 10 Grist 11 White 13 Saône
15 Refractometer 16 Lower 19 Odour 22 Apart 23 Gamay 24 Hides 25 Ranks
26 Rerun *Down* 1 As new 2 Rossi 3 Lodge 4 Apricot Brandy 5 Malts 6 Tinto
7 Lodge 12 Terre 14 Amend 16 Lager 17 Women 18 Rayas 19 Other 20 Older
21 Resin

Crossword 2
Across 1 St Vincent's Day 8 Inn 9 Weisswein 10 Ale drain 11 Fumé 14 A par round
17 Busy 18 Implores 20 Orange gin 22 IPA (India pale ale) 23 Beerenauslese
Down 1 Spiral 2 Vin de pays 3 New areas 4 Emilia Romagna 5 Test 6 Due 7 Yonder
12 Under ripe 13 Full ones 15 Absorb 16 Estate 19 Ogre 21 Ale

Crossword 3
Across 6 Lacryma Christi 9 Endear 10 Kabinett 11 By the sun 13 Avenue 15 Splash
17 Aspect 19 France 20 Italy DOC 22 Given two 24 Blames 26 Classification
Down 1 Blandy's Sercial 2 Acre 3 Cyprus 4 Chablais 5 Kiln 7 Asking
8 To trust someone 12 Halon 14 Every 16 Sweet gin 18 Midori 21 AOB Bar 23 Easy
25 Acid

Section eleven   Multiple choice questions

Part 1   A service industry
 1 d   2 b   3 a   4 b

Part 2   Wine from many countries

1 a	2 a	3 c	4 a	5 a	6 c	7 a	8 c	9 a	10 b	11 b
12 c	13 d	14 a	15 d	16 a	17 b	18 c	19 a	20 b	21 d	22 c
23 a	24 d	25 b	26 b	27 b	28 a	29 c	30 a	31 c	32 b	33 c
34 a	35 c	36 a	37 b	38 d	39 b	40 b	41 a	42 d	43 c	44 a
45 b	46 c	47 c	48 c	49 a	50 c	51 a	52 d	53 a	54 c	55 b
56 c	57 b	58 a	59 c	60 a	61 b	62 b	63 b	64 a	65 c	66 c
67 a	68 b	69 d	70 a	71 c	72 b	73 b	74 d	75 a	76 c	77 c
78 c	79 a	80 b	81 b	82 d	83 b	84 a	85 c	86 d	87 b	88 d
89 b	90 a	91 c	92 b	93 a	94 a	95 c	96 b	97 c	98 a	99 b
100 d	101 a	102 c	103 b	104 c	105 c	106 a	107 c	108 b	109 d	110 d
111 a										

Part 3 The service of wine

1 a	2 c	3 d	4 c	5 c	6 a	7 a	8 a	9 b	10 c	11 d
12 a	13 d	14 b	15 a	16 c	17 b					

Part 4 Fortified wine
1 b   2 a   3 d   4 c   5 b   6 d   7 a   8 d   9 c   10 c   11 a   12 c   13 b   14 a
15 a   16 c   17 a   18 d   19 b   20 c   21 d   22 b   23 a   24 d   25 c   26 b
27 b   28 b

Part 5   Eating and drinking
1 c   2 b   3 a   4 d   5 a

Part 6   Bar work
1 b   2 b   3 b   4 d   5 d   6 b   7 b   8 b   9 c   10 d

Part 7   Cellar work
1 a   2 b   3 b   4 a   5 d   6 b   7 b   8 c

Part 8   Brewing and distilling
1 b   2 a   3 d   4 b   5 a   6 b   7 c   8 a   9 b   10 b   11 d   12 b
13 b   14 a   15 d   16 a   17 c   18 d   19 b   20 c   21 a   22 b   23 b   24 c
25 a   26 b   27 c   28 a   29 c   30 d   31 a   32 d   33 c   34 a   35 d   36 a
37 c   38 c   39 b   40 d   41 b   42 c   43 b   44 b   45 a   46 c   47 b   48 a
49 d   50 b   51 b   52 b   53 a   54 d   55 a   56 c   57 d   58 b   59 a   60 b
61 d   62 c   63 d   64 c   65 d   66 b   67 b   68 b   69 b   70 b   71 c

Part 9   Storage of alcohol and other items
1 b   2 b   3 a   4 d   5 c   6 d   7 a   8 b   9 b   10 d

Part 10   Customer relations
1 b   2 c

Part 11   The law, health and safety
1 c   2 b   3 b   4 d   5 c   6 c   7 b   8 d   9 a   10 b   11 c   12 d   13 a   14 c